NCRP REPORT No. 151

Structural Shielding Design and Evaluation for Megavoltage X- and Gamma-Ray Radiotherapy Facilities

Recommendations of the
NATIONAL COUNCIL ON RADIATION PROTECTION AND MEASUREMENTS

December 31, 2005

National Council on Radiation Protection and Measurements
7910 Woodmont Avenue, Suite 400/Bethesda, MD 20814-3095

LEGAL NOTICE

Disclaimer

Library of Congress Cataloging-in-Publication Data

National Council on Radiation Protection and Measurements.
Structural shielding design and evaluation for megavoltage x- and gamma-ray radiotherapy facilities : recommendations of the National Council on Radiation Protection and Measurements.
p. cm. — (NCRP report ; no. 151)
"Issued December 2005."
Includes bibliographical references and index.
ISBN-13: 978-0-929600-87-1
ISBN-10: 0-929600-87-8
1. Radiology, Medical—Equipment and supplies—Safety measures. 2. Shielding (Radiation) 3. Radiotherapy—Safety measures. 4. Gamma rays—Physiological effect. 5. Radiology, Medical—Safety measures. I. Title. II. Series.
RA975.5.R3S77 2005
616.07'572--dc22

2006002275

[For detailed information on the availability of NCRP publications see page 227.]

Preface

This Report was developed under the auspices of Program Area Committee 2 of the National Council on Radiation Protection and Measurements (NCRP), the Committee that is concerned with operational radiation safety. The Report addresses the structural shielding design and evaluation for medical use of megavoltage x and gamma rays for radiotherapy and supersedes related material in NCRP Report No. 49, *Structural Shielding Design and Evaluation for Medical Use of X Rays and Gamma Rays of Energies Up to 10 MeV*, which was issued in September 1976. The descriptive information in NCRP Report No. 49 unique to x-ray therapy installations of less than 500 kV (Section 6.2) and brachytherapy (Section 7) is not included in this Report and that information in NCRP Report No. 49 for those categories is still applicable.

This Report was prepared through a joint effort of NCRP Scientific Committee 46-13 on Design of Facilities for Medical Radiation Therapy and the American Association of Physicists in Medicine (AAPM). NCRP gratefully acknowledges the support of AAPM for several of its members to serve on Scientific Committee 46-13 and the many opportunities that were made available for Scientific Committee 46-13 to meet at AAPM annual meetings. Serving on Scientific Committee 46-13 were:

James A. Deye, *Chairman*
National Cancer Institute
Rockville, Maryland

Vice-Chairmen

James E. Rodgers
Maryland Regional Cancer Care
Silver Spring, Maryland

Raymond K. Wu
OhioHealth Hospitals
Columbus, Ohio

Members

Peter J. Biggs
Massachusetts General Hospital
Boston, Massachusetts

Patton H. McGinley
Stone Mountain, Georgia

Richard C. McCall
Woodside, California

Liaisons

Kenneth R. Kase
Lyncean Technologies, Inc.
Palo Alto, California

Marc Edwards
Radiation Oncology Associates of Kansas City
Overland Park, Kansas

Consultants

Robert O. Gorson
Columbia, South Carolina

Jeffrey H. Kleck
Attania Corporation
San Jose, California

Nisy E. Ipe
San Carlos, California

NCRP Secretariat

Marvin Rosenstein, *Consultant*, 2001–2005
Eric E. Kearsley, *Staff Scientist / Consultant*, 1998–2001
James A. Spahn, Jr., *Senior Staff Scientist*, 1995–1998
Cindy L. O'Brien, *Managing Editor*
David A. Schauer, *Executive Director*

The contents are the sole responsibility of NCRP and do not necessarily represent the official views of the National Cancer Institute or the National Institutes of Health.

The Council wishes to express its appreciation to the Committee members, liaisons and consultants for the time and effort devoted to the preparation of this Report.

Thomas S. Tenforde
President

Contents

1. Introduction

1.1 Purpose and Scope

The purpose of radiation shielding is to limit radiation exposures to members of the public and employees to an acceptable level. This Report presents recommendations and technical information related to the design and installation of structural shielding for megavoltage x- and gamma-ray radiotherapy facilities. This information supersedes the recommendations in the National Council on Radiation Protection and Measurements' (NCRP) Report No. 49 (NCRP, 1976) pertaining to such medical radiotherapy facilities. Since the publication of NCRP Report No. 49, many facilities have been designed for accelerating voltages greater than the 10 MV maximum that was covered in that report.[1] Hence recent designs have had to refer to NCRP Report No. 51 (NCRP, 1977) and NCRP Report No. 79 (NCRP, 1984) in order to account for the higher accelerating voltages and the concomitant production of neutrons. In addition, the use of barriers constructed with composite materials has become commonplace.

This Report includes the necessary information for these higher accelerating voltages as well as a discussion of the various factors to be considered in the selection of appropriate shielding materials and in the calculation and evaluation of barrier thicknesses (Sections 1 through 6). Section 7 presents an extensive set of sample calculations, Appendices A and B provide supporting data figures and tables, respectively, and Appendix C discusses neutron monitoring for radiotherapy facilities.

This Report is mainly intended for those individuals who specialize in radiation protection, but it will also be of interest to architects, hospital administrators, and related professionals concerned with the planning of new radiotherapy facilities.

[1]Throughout this Report, "MV" will be used when referring to accelerating voltages and the endpoint energy of a bremsstrahlung spectrum, while "MeV" will be used when referring to monoenergetic photons or electrons.

Terms and symbols used in this Report are defined in the text and in the Glossary. Recommendations throughout this Report are expressed in terms of *shall* and *should* (in italics) where:

- *shall* indicates a recommendation that is necessary to meet the currently accepted standards of radiation protection; and
- *should* indicates an advisory recommendation that is to be applied when practicable or practical (*e.g.*, cost effective).

1.2 Quantities and Units[2]

The quantity recommended in this Report for shielding design calculations when neutrons, as well as photons, are present is dose equivalent (H). Dose equivalent is defined as the product of the quality factor for a particular type of ionizing radiation and the absorbed dose (D) [in gray (Gy)] from that type of radiation at a point in tissue (ICRU, 1993). The units of dose equivalent are J kg^{-1} with the special name sievert (Sv).

The quantity kerma measured in air (K_a) is recommended for shielding design calculations in low linear-energy-transfer (LET) environments (NCRP, 2004). For the direct measurement of shielding design quantities outside of treatment rooms in which photon or electron sources below 10 MeV are to be used (and therefore a significant yield of neutrons is not present), the result from an instrument calibrated for exposure [in roentgen (R)] is divided by 114 to obtain air kerma (K_a) (in gray), or by 104 to obtain an acceptable approximation for absorbed dose (in gray) or dose equivalent (in sievert) at a point in tissue.

Dose equivalent is used in this Report, where the quality factor has been assigned the value of unity for low-LET radiation (*i.e.*, photons and electrons) (NCRP, 1993). For facilities likely to produce a significant yield of neutrons (*e.g.*, sources producing photons or electrons with energies above 10 MeV) (NCRP, 1984; Shultis and Faw, 1996), the direct determination of the dose equivalent from all components of the radiation field (*i.e.*, low- and high-LET radiations) needs to be made by measurement or calculation.

Figure 1.1 (McDonald *et al.*, 1998) presents fluence-to-dose equivalent conversion coefficients for neutrons as a function of neutron energy, as compiled over the last four decades. Despite many

[2]See Glossary for definitions of the quantities and units.

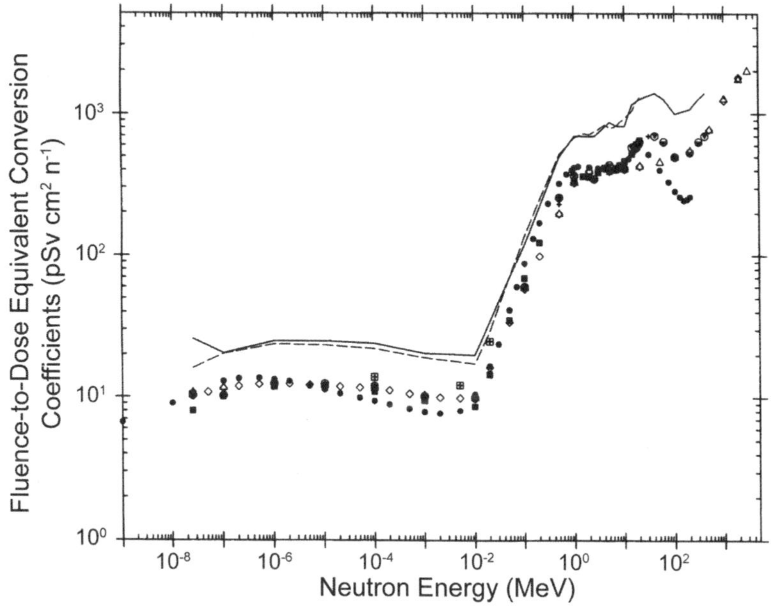

Fig. 1.1. Fluence-to-dose equivalent conversion coefficients (McDonald *et al.*, 1998). [● (ICRP, 1996; ICRU, 1998), ▼ (DOE, 1993; NRC, 1996), — (NCRP, 1991), - - and ■ (before Paris) (ICRP, 1987), △ (ICRP, 1973), ◊ (Patterson and Thomas, 1973), O (NCRP, 1971), □ (ICRP, 1964), and + (NBS, 1960).

refinements that have been made in these fluence-to-dose equivalent conversion coefficients, it is clear that practical radiation protection for neutrons [typically conducted within ±35 % uncertainty at the 95 % confidence level for dose-equivalent rates <0.02 mSv h^{-1} (NCRP, 1991)] will not be significantly improved by reducing the differences among the data sets of conversion coefficients in Figure 1.1 (on the order of 10 to 30 % for neutrons <20 MeV). Thus, most conversion coefficients in the literature for neutrons can be used directly. Further discussion of neutron monitoring equipment and its application to determination of neutron dose equivalent for radiotherapy facilities is provided in Appendix C.

The recommended radiation protection quantity for the limitation of exposure to people from sources of radiation is effective dose (E), defined as the sum of the weighted equivalent doses to specific organs or tissues (*i.e.*, each equivalent dose is weighted by the corresponding tissue weighting factor for the organ or tissue) (NCRP, 1993). The equivalent dose to a specific organ is obtained by weighting the mean absorbed dose in a tissue or organ by a radiation weighting factor.[3]

Application of dose equivalent and effective dose in this Report is described in Section 1.4.

NCRP has adopted the use of the International System (SI) of Units in its publications (NCRP, 1985). In addition, this Report will occasionally utilize both SI and non-SI units to describe certain characteristics for building materials, since non-SI units are in common use in the architectural community in the United States.

1.3 Controlled and Uncontrolled Areas

A controlled area is a limited-access area in which the occupational exposure of personnel to radiation or radioactive material is under the supervision of an individual in charge of radiation protection. This implies that access, occupancy and working conditions are controlled for radiation protection purposes. In radiotherapy facilities, these areas are usually in the immediate areas where radiation is used, such as treatment rooms and control consoles, or other areas that require control of access, occupancy and working conditions for radiation protection purposes. The workers in these areas are those individuals who are specifically trained in the use of ionizing radiation and whose radiation exposure is usually individually monitored.

Uncontrolled areas for radiation protection purposes are all other areas in the hospital or clinic and the surrounding environs. Note that trained radiation oncology personnel and other trained workers, as well as members of the public, frequent many areas near controlled areas such as examination rooms or restrooms. These areas are treated as uncontrolled in this Report. The choice of appropriate occupancy factors ensures the protection of both those who are occupationally exposed as well as others who might be exposed in these areas.

[3]The radiation weighting factor (in effective dose) and the quality factor (in dose equivalent) are both intended to take into account the differences, relative to photons, in the effectiveness in inducing stochastic effects at low absorbed doses for different types of ionizing radiation. The numerical value of the quality factor is determined by the values of the stopping powers for the spectrum of the charged particles at the point in tissue where the energy is absorbed. The numerical value of the radiation weighting factor is assigned according to the type and energy of the ionizing radiation that is incident on the body.

1.4 Shielding Design Goals and Effective Dose

In this Report, shielding design goals (*P*) are levels of dose equivalent (*H*) used in the design calculations and evaluation of barriers constructed for the protection of workers or members of the public. There are different shielding design goals for controlled and uncontrolled areas. The approach for structural shielding design for radiotherapy facilities and the application of shielding design goals and the NCRP recommended effective dose (*E*) limits for workers and members of the public, as they apply to controlled and uncontrolled areas in the design of new facilities, is discussed in this Section.

It is not practical to base shielding design directly on *E*. Determination of *E* is complex, and depends on the attenuation of photons and neutrons in the body in penetrating to the radiosensitive organs and hence on the energy spectra of the photons and neutrons, and also on the posture of the recipient with respect to the source. Rotational exposure is most likely, since it is probable that an individual is moving about and would not be exposed from one direction only. For the purposes of this Report, the shielding design goals are stated in terms of *H* (in millisievert) at the point of nearest occupancy beyond the barrier. For example, the distance of closest approach to a wall bounding a radiotherapy room can be assumed to be not less than 0.3 m.

Shielding design goals (*P*) are practical values, for a single radiotherapy source or set of sources, that are evaluated at a reference point beyond a protective barrier. When used in conjunction with the conservatively safe assumptions in this Report, the shielding design goals will ensure that the respective annual values for *E* recommended in NCRP Report No. 147 (NCRP, 2004) and in this Report for controlled and uncontrolled areas are not exceeded. Shielding design goals are expressed most often as weekly values since the workload (Section 1.5) for a radiotherapy source has traditionally utilized a weekly format.

1.4.1 *Controlled Areas*

The employees who work in controlled areas have significant potential for exposure to radiation in the course of their assignments, or are directly responsible for or involved with the use and control of radiation. Generally, these employees have training in radiation management and are subject to routine personal monitoring.

NCRP recommends an annual limit for E for these individuals of 50 mSv y^{-1} with the cumulative E not to exceed the product of 10 mSv and the worker's age in years (exclusive of medical and natural background radiation) (NCRP, 1993). That notwithstanding, NCRP (1993) recommends that for design of new facilities, E *should* be a fraction of the 10 mSv y^{-1} implied by the cumulative effective dose limit. Another consideration is that a pregnant radiation worker *should not* be exposed to levels that result in greater than the monthly equivalent-dose (H_T) limit of 0.5 mSv to the worker's embryo or fetus (NCRP, 1993). To achieve both recommendations, this Report recommends a fraction of one-half of that E value, or 5 mSv y^{-1}, and a weekly shielding design goal (P) of 0.1 mSv dose equivalent (H) (*i.e.*, an annual H value of 5 mSv) for controlled areas. The P value adopted in this Report would allow pregnant radiation workers continued access to their work areas.

Recommendation for Controlled Areas:
Shielding design goal (P) (in dose equivalent): 0.1 mSv week^{-1} (5 mSv y^{-1})

1.4.2 *Uncontrolled Areas*

Uncontrolled areas are those occupied by individuals such as patients, visitors to the facility (*e.g.*, patient visitors, delivery service representatives, and consultants), and employees who do not work routinely with or around radiation sources. Areas adjacent to, but not part of, the radiotherapy facility are also uncontrolled areas.

Based on ICRP (1991) and NCRP (1993) recommendations for the annual limit of effective dose to a member of the public, shielding designs *shall* limit exposure of all individuals in uncontrolled areas to an effective dose that does not exceed 1 mSv y^{-1}. After a review of the application of the guidance in NCRP (1993) to medical radiation facilities, NCRP concluded that a suitable source control for shielding individuals in uncontrolled areas in or near medical radiation facilities is an effective dose of 1 mSv in any year (NCRP, 2004). This recommendation can be achieved for the medical radiation facilities covered in this Report with a weekly shielding design goal (P) of 0.02 mSv dose equivalent (H) (*i.e.*, an annual H value of 1 mSv) for uncontrolled areas.

Recommendation for Uncontrolled Areas:
Shielding design goal (P) (in dose equivalent): 0.02 mSv week^{-1} (1 mSv y^{-1})

1.4.3 *Shielding Design Assumptions*

A radiotherapy facility that utilizes the *P* values given above would produce *E* values lower than the recommendations for *E* in NCRP (2004) and this Report for controlled and uncontrolled areas. This is the result of the conservatively safe nature of the shielding design methodology recommended in this Report. Several examples of this conservatism, and the impact of each, are given below.

- Attenuation of the primary beam by the patient is neglected. The patient typically attenuates the primary beam by 30 % or more.
- The calculations of recommended barrier thickness often assume perpendicular incidence of the radiation. If not assumed, the effect would vary in magnitude, but would always be a reduction in the transmission through the barrier for photons and neutrons that have nonperpendicular incidence. This is due to both the slant thickness of the barrier as well as the increased distance to the barrier.
- Leakage radiation from radiotherapy equipment is assumed to be at the maximum value recommended by IEC (2002) for the radiotherapy device, although in practice the leakage radiation is often less than this value. If the maximum value were not assumed, the effect would be a reduction in leakage radiation and its contribution to secondary radiation.
- The recommended occupancy factors for uncontrolled areas are conservatively high. For example, very few people spend 100 % of their time in their office. If more realistic occupancy factors were used, the effect would vary in magnitude, but would generally result in a reduction in the amount of exposure received by an individual located in an uncontrolled area.
- The minimum distance to the occupied area from a shielded wall is assumed to be 0.3 m. This is typically a conservatively safe estimate for most walls and especially for doors. If a value >0.3 m were assumed, the effect would vary, but radiation levels decrease with increasing distance (with a possible exception for unusual situations such as skyshine through thin ceilings).
- Often, when data are hard to estimate, such as in the design of accelerator facilities that will employ special procedures, safety factors are recommended (for example, multiplication by 1.5 in Section 3.2.5.4).

- The "two-source rule" (*i.e.*, the procedure when more than one source is involved) (Glossary) is applied whenever separate radiation components are combined to arrive at a barrier thickness. This has been shown to be a conservatively safe assumption since the tenth-value layer (*TVL*) and half-value layer (*HVL*) of the more penetrating radiation is always used. The two-source rule is even more conservatively safe when applied to dual-energy machines, even though the individual energies cannot be used simultaneously.

The conservatively safe factors discussed above will give a significant measure of assurance to the shielding designer that the actual dose equivalent transmitted through a barrier designed with the methodology given in this Report will be much less than the applicable shielding design goal. A new facility can be designed using the methodology in this Report without a significant increase in the cost or amount of structural shielding previously required.

1.4.4 *Measurements to Assess Compliance with the Shielding Design Goals*

For practical reasons, measurements made to assess the adequacy of barriers are generally made over periods of time that are much less than the length of time (*i.e.*, weekly or annually) specified in the recommended shielding design goals in this Report for controlled and uncontrolled areas. In this Report, the period of time most convenient for shielding calculations is one week. Accordingly, instantaneous or near-instantaneous measurements of the dose-equivalent rate are only appropriate in the determination of compliance with the shielding design goals if appropriate allowances are made for all of the factors that influence the projected weekly dose equivalent at the appropriate location behind the barrier. It is the weekly dose equivalent that is used to determine compliance with the shielding design goals.

1.5 Workload

The workload (*W*) for radiotherapy equipment covered in this Report is the time integral of the absorbed-dose rate determined at the depth of the maximum absorbed dose, 1 m from the source. The most common period of time over which *W* is specified is one week.

The units for W are Gy week^{-1} and conversion to a workload W_2 at a distance d_2 different than 1 m would be $W_2 = W\,(1\ \text{m})^2 / (d_2)^2$.

The value for W is usually specified as the absorbed dose from photons delivered to the isocenter in a week, and is selected for each accelerator based on its projected use. This is usually estimated from the average number of patients (or fields) treated in a week and the absorbed dose delivered per patient (or field). It should also include an estimate of the average weekly absorbed dose delivered during quality control checks, calibrations or other physics measurements.

Treatments on modern clinical accelerators often involve using low- and high-energy x-ray beams[4] and electron beams of various energies. For dual-energy machines, the workload at the higher energy will usually determine the shielding requirement. However, in some situations, to determine the required barrier thicknesses for both primary and secondary radiations it may be necessary to consider separately the workloads for each x-ray beam quality. Workload for electron beam operation can be disregarded, except for shielding accelerators with electron beam-only operation, such as dedicated intraoperative facilities (Section 5.8). Modern radiotherapy facilities often employ techniques, such as intensity modulated radiation therapy (IMRT), that build up an absorbed-dose distribution in the target volume through the accumulation of multiple beamlets (small area beams). These treatment delivery methods can lead to leakage-radiation workloads that are significantly greater than the total absorbed dose at the isocenter, and this has led various authors to create a workload efficiency factor (Mutic and Low, 1998; Rodgers, 2001) as well as to decouple the primary and secondary beam workloads (Rodgers, 2001). These concepts are employed in this Report and dealt with in detail in Section 3.

1.6 Use Factor

The use factor (U) is the fraction of a primary-beam workload that is directed toward a given primary barrier. The value for U will depend on the type of radiation installation. For example, a traditional facility with a beam that rotates about an isocenter will usually have a symmetric distribution of gantry treatment angles and

[4]For purposes of shielding design for x-ray beams, the terms low-energy accelerator (defined as ≤10 MV accelerating voltage) and high-energy accelerator (defined as >10 MV accelerating voltage) will be employed.

these will be predominately in the four primary angles (0, 90, 180 and 270 degrees). However, a facility that is used for total-body irradiation (TBI) with the patient at an extended distance will have a large use factor for the direction of the TBI treatments. And facilities that perform a large number of tangential breast treatments will have significant use factors for oblique room angles. Therefore, the actual methods of patient treatments *shall* be considered when determining the design use factors as further discussed in Section 3.

1.7 Occupancy Factor

The occupancy factor (T) for an area is the average fraction of time that the maximally exposed individual is present while the beam is on. Assuming that use of a radiotherapy unit is relatively uniformly spread out over the workweek, the occupancy factor is the fraction of the working hours in the week that this individual would occupy the area, averaged over the year. For example, an uncontrolled area adjacent to a treatment room having an assigned occupancy factor of 1/40 would imply that the maximally exposed individual would spend an average of 1 h week^{-1} in that area every workweek for a year. The occupancy factor for an area is not the fraction of time that it is occupied by any persons, but rather it is the fraction of the time it is occupied by the single person who spends the most time there. Thus a waiting room might be occupied at all times during the working day, but have a very low occupancy factor since no single person is likely to spend more than 50 h y^{-1} in a given waiting room. Occupancy factors in uncontrolled areas will rarely be determined by visitors to the facility or its environs who might be there only for a small fraction of a year. The maximally exposed individual will normally be an employee of the facility.

The occupancy factor for controlled areas is usually assigned a value of unity. However, there can be situations in which access to a controlled area is restricted even for radiation workers when radiation is being produced (*e.g.*, an accelerator equipment support room). In such cases, the qualified expert designing the shielding requirements for that controlled area may use local information on occupancy of the area. An example is provided in Section 7.2.3. Conversely, if low occupancy factors are used, the instantaneous and time-averaged dose-equivalent rates (Section 3.3) can become quite high and require careful consideration of the occupancy information, to ensure that the shielding design goal for controlled areas is not exceeded.

In some cases, a clinic may plan to operate equipment longer than a normal 40 h workweek. In this case, the occupancy factor *shall* be determined by the ratio of the average time the maximally exposed individual will be present to the total average time that the equipment is used during the week. The period over which the average *shall* be estimated is 1 y.

1.8 Protective Barriers

In radiotherapeutic applications, the radiation consists of primary and secondary radiations (Figure 1.2). Primary radiation, also called the useful beam, is radiation emitted directly from the equipment that is used for patient therapy. A primary barrier is a wall, ceiling, floor or other structure that will intercept radiation emitted directly from the source. It needs to attenuate the useful beam and also any secondary radiation that impinges on it to the appropriate shielding design goal. Secondary radiation consists of radiation scattered from or produced by interactions with the patient and other objects as well as the leakage radiation from the protective housing of the source. A secondary barrier is a wall, ceiling floor or other structure that will intercept the secondary radiation. It needs to attenuate the secondary radiation to the appropriate shielding design goal. A full discussion of primary and secondary barriers is given in Section 2.

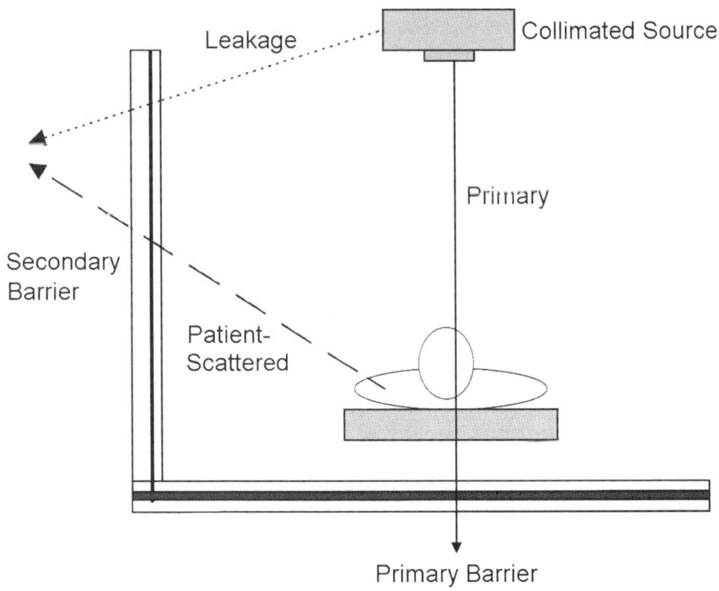

Fig. 1.2. Schematic of radiation sources (primary, leakage and patient-scattered) and the primary and secondary barriers.

1.9 Basic Principles

Exposure of individuals to primary and secondary radiations can be reduced by one or a combination of the following methods:

- increasing the distance between the individual and the sources of the radiation,
- limiting the exposure time, and
- interposing protective shielding between the individual and the radiation sources.

The dose rate from the source varies inversely as the square of the distance from the source. It is usually assumed that the individual to be protected is at least 0.3 m from the barrier. The exposure time involves both the time that the radiation beam is on and the fraction of the beam-on time during which a person is in the radiation field. In addition to time, distance and shielding, administrative controls such as limiting access to an area or additional surveillance can be, and often are, used to reduce or avoid exposure to ionizing radiation (NCRP, 1990; 1993).

1.10 General Concepts

The term qualified expert used in this Report is defined as a medical physicist or a health physicist who is competent to design radiation shielding in radiotherapy facilities, and who is certified by the American Board of Radiology, American Board of Medical Physics, American Board of Health Physics, or Canadian College of Physicists in Medicine.

Radiation shielding *shall* be designed by a qualified expert to ensure that the required degree of protection is achieved. The qualified expert *should* be consulted during the early planning stages since the shielding requirements may affect the choice of location of radiation facilities and type of building construction. The qualified expert *should* be provided with all pertinent information regarding the proposed radiation equipment and its use, type of building construction, and occupancy of nearby areas. It may also be necessary to submit the final shielding drawings and specifications to pertinent regulatory agencies for review prior to construction. Other aspects of radiotherapy facility design, such as interlocks, warning signs, warning lights, electrical safety, and room lighting are mentioned in this Report, but these aspects do not represent a complete treatment of these topics. Further considerations of these topics can be found in NCRP Report No. 102 (NCRP, 1989).

The shielding of the radiotherapy room *shall* be so constructed that the protection is not compromised by joints, by openings for ducts, pipes or other objects passing through the barriers, or by conduits, service boxes, or other structural elements embedded in the barriers. Door design for high-energy machines also requires special consideration to ensure adequate protection without sacrificing operational efficiency.

There is considerable variation in the shielding requirements for radiotherapy installations owing to the wide range of energies and different types of equipment and clinical techniques used. Careful planning may result in appreciable savings, particularly in the high-energy range where shielding is very costly. Provision for future requirements may prevent expensive alterations.

The shielding design goals (*P* values) in this Report apply only to new facilities and new construction and will not require retrofitting of existing facilities. This Report is intended for use in planning and designing new facilities and in remodeling existing facilities. Facilities designed before the publication of this Report and meeting the requirements of NCRP Report No. 49 (NCRP, 1976) need not be reevaluated (NCRP, 1993) unless there are changes in a facility's design or use. New equipment, significant changes in the use of equipment, or other changes that may have an impact on radiation protection of the staff or members of the public require an evaluation by a qualified expert.

Since corrections or additions after facilities are completed are expensive, it is important that structural shielding be properly designed and installed in the original construction process. It is also advisable that the planning includes consideration of possible future needs for new equipment and changes in practice or use, increased workloads, and changes in the occupancy of adjacent spaces.

The final drawings and specifications *should* be reviewed by the qualified expert and by the pertinent federal, state or local agency if applicable, before construction is begun. Because any radiation exposure may have an associated level of risk (NCRP, 1993), it is important that the qualified expert review the completed facility design to ensure that all anticipated exposures also are consistent with the ALARA (as low as reasonably achievable) principle (NCRP, 1990; 1993) (see Glossary). The cost of increasing shielding beyond the minimum value often represents only a small increase in cost.

It is often impractical to make an overall experimental determination of the adequacy of the shielding prior to the completion of the building construction and the installation of the radiation equipment. Periodic inspections during the entire construction period *should* be performed. Sometimes properly constructed shielding is compromised by subsequent changes that are made to install ducts, recessed boxes, or other hardware. These alterations could be made to walls, ceilings, or floors. Hence, there *should* be periodic checks of the continued validity of shielding assumptions and the integrity of the barriers.

After construction, a performance assessment (*i.e.*, a radiation survey), including measurements in controlled and uncontrolled areas, *shall* be made by a qualified expert to confirm that the shielding provided will achieve the respective shielding design goal (*P*). The performance assessment is an independent check that the assumptions used in the shielding design are conservatively safe. In addition, it is good radiation protection practice to monitor periodically to ensure that the respective recommendations for *E* (Sections 1.4.1 and 1.4.2) continue to be met during facility operation.

This Report does not attempt to summarize the regulatory or licensing requirements of the various authorities that may have jurisdiction over matters addressed in this Report. Similarly, no recommendations are made on administrative controls that site operators may choose to implement. It is expected that the qualified expert will be fully aware of these matters and account for them in the final shielding design.

While specific recommendations on shielding design methods are given in this Report, alternate methods may prove equally satisfactory in providing radiation protection. The final assessment of the adequacy of the design and construction of protective shielding can only be based on the post-construction survey performed by a qualified expert. If the survey indicates shielding inadequacy, additional shielding or modifications of equipment and procedures *shall* be made.

1.11 Types of Radiotherapy Installations

Modern radiation therapy employs an array of treatment techniques that have resulted from a better understanding of disease processes and advanced imaging and radiation delivery technologies. These employ both hardware and software that allow for real-time imaging of the target anatomy and dynamic modification of the shape and intensity of the radiation fields. Thus many facilities now utilize TBI (AAPM, 1986a), IMRT (Purdy *et al.*, 2001), stereotactic radiosurgery (SRS) and stereotactic radiotherapy (SRT) (AAPM, 1995). These techniques often result in significant changes in the workload and use factors for the facility when compared to the conventional treatment methodologies. A smaller number of facilities, at this time, use intraoperative radiotherapy (IORT) (Palta *et al.*, 1995); and, since this involves the exclusive use of electron beams, many of the methods proposed in this Report can be used for the shielding design of such a facility.

In total-body photon irradiations, the maximum field size directed onto a specific primary barrier is often used with beam-on times of 15 min or more (AAPM, 1986a). Thus, the use factor for that barrier can be much larger than would be the case for routine fields delivered to the patient from multiple directions.

Intensity modulated radiation therapy (IMRT) can be accomplished with different technologies, but the net result is that the standard therapeutic absorbed dose is delivered from many directions around the patient with as much as 10 times the normal beam-on time (Purdy *et al.*, 2001). This results because much of the radiation produced in the treatment machine head is attenuated by the collimator before it reaches the patient. In this situation, the fluence on the primary barriers is quite similar to the conventional treatment regimen but the leakage radiation on the secondary barriers may be much larger.

With SRS and SRT, high individual absorbed doses are delivered to patients and therefore both the primary and secondary-barrier workloads can be greater than in the standard case (AAPM, 1995). Likewise, multiple, oblique angles are used and this can skew assumptions about the use factors for the barriers if they were not explicitly considered in the design.

The actual effect of any of these special situations on a weekly or yearly basis may be offset by the fact that the patient and machine setup times are also significantly larger than in conventional treatments and therefore fewer such TBI, IMRT, or SRS treatments can be delivered over the course of the workday than conventional treatments. Facilities that anticipate the use of one or more of these advanced techniques should carefully evaluate their anticipated weekly workload.

1.12 Strategic Shielding Planning

Strategic shielding planning for a radiotherapy facility incorporates a knowledge of basic planning and shielding principles. The strategic planning concept involves the use of shielding options dictated by a knowledge of the sources of radiation in a facility, the occupancy and usage of adjacent areas, and whether specific walls, floors and ceilings must be considered primary or secondary barriers.

The qualified expert and architect need to be aware, for example, that the use of exterior walls and adjacent spaces, both horizontal and vertical, can often be cost effective elements in the design of radiation shielding. For example, a corridor can be used to separate offices and support rooms from the treatment rooms rather than leaving these rooms adjacent to one another. This strategy will often reduce the amount of required shielding to protect the office occupants. The corridor is a low occupancy area and the occupied spaces (offices and lounges) are at least 2.5 m further from the source of radiation, though they may still be the determining factor for the barrier thickness. The same strategy applies for spaces above and below; locating a treatment room below a corridor or mechanical room rather than an occupied office is an effective strategy for reducing shielding requirements.

The effective and efficient use of shielding materials and the development of optimal design strategies require communication and cooperation among the architect, facility representative, and qualified expert.

The project development process will vary from institution to institution. In addition, small projects may be developed differently from large projects. However, a project development process will most likely consist of the following five phases.

1.12.1 *Planning and Budgeting*

Almost every institution or business goes through an annual budgeting process. In addition, most institutions will undertake major strategic planning sessions every few years. During the budgeting process or strategic planning process, decisions will be made to enter into new or existing businesses or services, or to purchase new capital equipment. When these processes involve new construction or purchase of new radiation producing equipment, the qualified expert *should* be consulted to help develop comprehensive budgets and schedules. While the cost of shielding is a relatively modest component of any project cost, the goal is to be as accurate as possible in the initial decision-making process.

1.12.2 *Programming*

The purpose of the programming phase is to prepare a detailed comprehensive list of rooms, their sizes, and any special requirements of each room. During this phase the qualified expert can provide information concerning shielding requirements and suggest floor plans that will help minimize shielding requirements.

Cooperation between the qualified expert and the space programmer at this phase will help create a safe, efficient health care environment.

1.12.3 *Schematic (Preliminary) Design*

During the schematic or preliminary design phase the architect begins to organize the rooms into a workable and efficient plan to illustrate the scope of the project. Single-line floor plans to scale, notes and outline specifications of major materials and systems are produced. The qualified expert *should* be involved in the schematic design phase. The qualified expert can help determine appropriate floor plans and point out walls, floors and ceilings which will need to be studied for potential shielding requirements. The architect and qualified expert can begin to consider appropriate materials and systems that will meet project goals and contribute to the shielding design.

1.12.4 *Design Development*

In this phase, rooms, sizes and locations will be determined in much greater detail and the design will be finalized. The architect and mechanical, electrical, plumbing and structural engineers will begin to fix the scope of work. Structural systems and major duct sizing and location will be determined. The qualified expert *should* be provided with the proposed layout for each room in order to determine which walls, floors or ceiling will contain conduits, and heating, ventilation and air-conditioning (HVAC) and high-voltage ducts. At this point, the qualified expert can work with the architect and structural engineer to become aware of the actual structural systems to be used and the design thickness of floor and roof slabs. In renovation projects, architects and engineers will investigate as-built conditions including types of existing structural systems, and floor and roof slab thicknesses. It is important for the qualified expert and the architect to also determine the occupancy of the spaces above and below the treatment room.

1.12.5 *Construction Document Preparation*

Construction documents, contract documents, working drawings, and blueprints are almost interchangeable terms used to identify the drawings and specifications prepared during this phase. At this point, details of the project are finalized. Dimensions, floor plans, wall sections, wall elevations, system details, materials, and construction directions are documented. This set of documents illustrates the detail drawings such as door frames, wall penetrations, and any of the shielding details required to meet the qualified expert's requirements. The location and size of vertical duct chases are shown on the drawings and the shielding specifications are detailed in the wall and floor sections. The qualified expert *should* review the construction documents with the architect prior to the release of the documents for bidding. The qualified expert *shall* specify where shielding is needed and the amount of shielding (including type and density of material) required prior to construction. In addition, the qualified expert *shall* review and approve any final changes that may modify shielding requirements.

1.12.6 *Construction Inspection*

It is recommended that the qualified expert carry out a physical inspection of the facility during construction. The inspection *should* include an evaluation of at least the following items:

- thickness and density of concrete;
- thickness of metal shielding and polyethylene used for neutron shielding;
- thickness of metal behind recesses in the concrete (*e.g.*, laser boxes);
- HVAC shielding baffle (Section 4.4) if used;
- location and size of conduit or pipe used for electrical cable of any type; and
- verification that the shielding design has been followed.

A summary document outlining the results of the construction inspection *shall* be prepared by the qualified expert and forwarded, as appropriate, to the owner of the facility, the architectural firm involved in the construction and the governing regulatory agency. Any items of noncompliance *shall* be clearly indicated and recommendations for corrections should be made.

1.13 Documentation Requirements

The following documentation *shall* be maintained on a permanent basis by the operator of the facility:

- shielding design report, including assumptions and specifications;
- construction, or as-built, documents showing location and amounts of shielding material installed;
- post-construction survey reports;
- information regarding remedies, if any were required; and
- more recent reevaluations of the room shielding relative to changes (*e.g.*, in utilization) which have been made or are still under consideration.

2. Calculational Methods

2.1 Basic Concepts

Shielding design for medical radiation therapy facilities has been based on simple empirical equations developed by Mutscheller (1925; 1926) and later refined by NCRP (1976; 1977). The basic concept is depicted in Figure 2.1, in which an individual at Location O must be protected from a radiation source S that is a distance *d* away. The level of protection is given by the applicable shielding design goal (*P*) which depends on whether the person is occupationally exposed in a controlled area or a member of the public in an uncontrolled area. If the radiation source produces a level $>P$ at Location O, then a barrier B is used to attenuate the radiation level so that *P* is not exceeded. In keeping with the principle of ALARA, it also may be cost efficient to design to a value of dose equivalent that is less than the applicable value of *P* (NCRP, 1990).

Sources of radiation dealt with in this Report are bremsstrahlung photons produced by medical linear accelerators or gamma rays from isotopic machines and the secondary radiations produced by these photons as shown in Figure 2.2 (NCRP, 1984). The bremsstrahlung process is confined to the target of the accelerator while the photoneutron production process (γ,n) occurs in both the accelerator head (Mao *et al.*, 1997) and the room shielding. Neutron

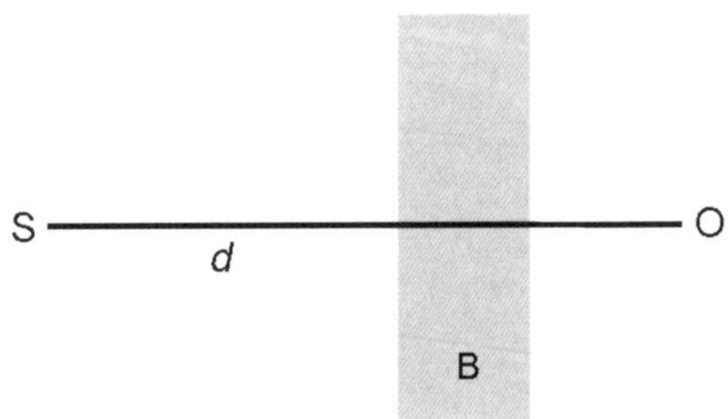

Fig. 2.1. Basic shielding schematic of an individual at Location O protected from radiation source S at *d* distance away by a shield at B.

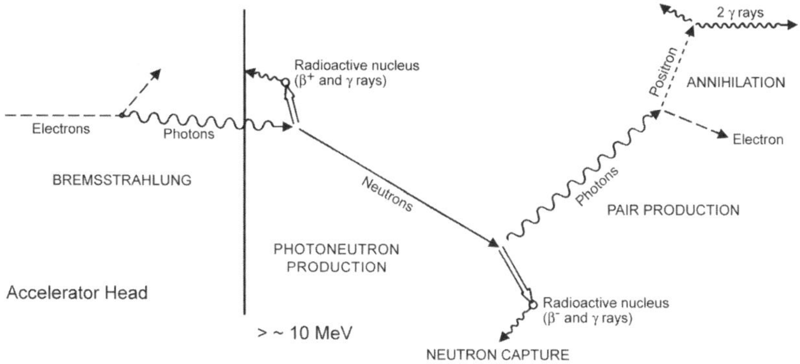

Fig. 2.2. Production of radiation types in a linear accelerator. Radiations to the right of the line have significant production cross sections in accelerators with photon energies above ~10 MeV.

capture gamma rays are largely confined to production in the room shielding and result from photoneutrons produced when the primary photons have energies above the neutron binding energy of roughly 8 MeV for most nuclides. In fact, as shown in NCRP Report No. 79 (Figure 15 in NCRP, 1984), neutron yields from most electron linear accelerator materials do not become significant until the incident energy exceeds 10 MeV. As pointed out by Shultis and Faw (1996), the photoneutron production cross section increases with photon energy by several orders of magnitude to a broad maximum at photon energies of ~20 to 23 MeV for light nuclei (atomic number less than ~40) and 13 to 18 MeV for medium to heavy nuclei. Therefore, for shielding of accelerators with accelerating voltages of 10 MV or less, usually only photons and electrons need to be addressed, although there can be exceptional situations with room shielding consisting of high-Z material such as lead and steel only, or laminated barriers with insufficient hydrogenous material.

As discussed in Section 1.8, two types of radiation barriers are routinely considered: primary and secondary. The primary barrier can be irradiated directly by photons from the target or source, while the secondary barrier receives radiation resulting from scatter of the primary beam by the patient (*i.e.*, patient-scattered radiation) and/or the surfaces of the treatment room in addition to the radiation transmitted through the accelerator head (*i.e.*, head-leakage radiation). Secondary radiation is emitted in all directions and covers all of the treatment room surfaces. Primary radiation, however, is limited in direction by the placement of the accelerator in the treatment room and the maximum beam size. An adequately designed primary barrier will be more than sufficient as a barrier for all sources of secondary radiation. In the following, each type of barrier is considered separately.

2.2 Primary Barriers

2.2.1 *Standard Approach*

In the usual approach, primary barriers are designed to attenuate the photon beam emanating from the treatment unit that is directly incident on the barrier. The primary barrier is also expected to adequately attenuate the dose equivalent beyond the barrier that results from secondary products of the photon beam. One example of this is the photoneutrons produced by the primary beam in the accelerator head as well as within the primary barrier itself. If the empirical methods given below are followed, the photon barrier will also be adequate for the secondary neutrons and neutron capture gamma rays that can originate along the path of the primary beam (IAEA, 1979).

For an adequate barrier the ratio of the dose equivalent transmitted through the barrier to the shielding design goal (P) needs to be less than or equal to one. Hence the transmission factor of the primary barrier (B_{pri}) that will reduce the radiation field to an acceptable level is given by Equation 2.1.

$$B_{pri} = \frac{P\, d_{pri}^2}{WUT} \tag{2.1}$$

In Equation 2.1:[5]

P = shielding design goal (expressed as dose equivalent) beyond the barrier and is usually given for a weekly time frame (Sv week^{-1})

d_{pri} = distance from the x-ray target to the point protected (meters)

W = workload or photon absorbed dose delivered at 1 m from the x-ray target per week (Gy week^{-1})[6]

[5]All distances d in this Report are referenced to a distance of 1 m from the source of the designated radiation field. Therefore, due to the inverse square law it is understood that all d^2 values in the equations and calculations in this Report are divided by (1 m)^2.

[6]While W is expressed here in absorbed dose (Gy week^{-1}), that is equivalent to W being expressed in dose equivalent (Sv week^{-1}), since the quality factor for photons is assigned a value of unity. Therefore, the transmission factor (B) for photons is a unitless quantity.

U = use factor or fraction of the workload that the primary beam is directed at the barrier in question

T = occupancy factor for the protected location or fraction of the workweek that a person is present beyond the barrier. This location is usually assumed to be 0.3 m beyond the barrier in question (see Table B.1 in Appendix B for recommended occupancy values)

The thickness of the barrier can then be determined using tenth-value layers based on the energy of the accelerator and type of shielding material (see Figures A.1a and A.1b in Appendix A and Table B.2 in Appendix B). In this case, the required number (n) of TVLs is given by:

$$n = -\log(B_{pri}) \tag{2.2}$$

and the barrier thickness ($t_{barrier}$) is given by:

$$t_{barrier} = TVL_1 + (n-1)\ TVL_e \tag{2.3}$$

The first (TVL_1) and equilibrium (TVL_e) tenth-value layers of the desired material are used to account for the spectral changes in the radiation as it penetrates the barrier. Thus, when a barrier thickness (t) is greater than the first TVL_1, the total transmission factor (B) is given by Equation 2.4.

$$\begin{aligned} B &= (10^{-1})\ 10^{-\left[\frac{(t-TVL_1)}{TVL_e}\right]} \\ &= 10^{-\left\{1+\left[\frac{(t-TVL_1)}{TVL_e}\right]\right\}} \end{aligned} \tag{2.4}$$

If the material used in the primary barrier is concrete (whether ordinary or heavy; see Sections 4.3.1 and 4.3.2), then experience has shown that the barrier will adequately absorb all photoneutrons and neutron capture gamma rays and no additional barrier is required. This is due to the relatively high hydrogen content of concrete and its resultantly high neutron absorption cross section. If, on the other hand, materials other than concrete are used in the primary barrier, then special considerations are required and these are covered in Section 2.2.3.

Typical workloads are discussed in Section 3. Though modern techniques such as IMRT are known to require very large numbers of monitor units or beam-on time, these techniques may use very small beamlets (less than ~1 cm^2) for the primary beam and therefore the workload averaged over any 100 cm^2 area on the primary barriers is approximated by standard workloads.

Modern linear accelerators are also capable of very high instantaneous dose rates and these are reflected in any survey meter readings. However, it is noted that the shielding design goals are for a specific time period and hence other considerations (such as use and occupancy factors) need to be taken into account in deciding on the adequacy of any barrier. Survey meter readings may be used to assess cautionary levels that deserve further consideration but the final adequacy of the shielding *shall* be based upon compliance with the recommended shielding design goals, which are defined for a period of one week (Section 3.3).

In general, the primary-barrier thickness *should* be calculated for the perpendicularly incident beam and held constant over the whole barrier width. Not only does this ensure a conservatively safe thickness for the barrier but, importantly, it also adds to the quality assurance (QA) of the construction, since uniform thicknesses are more reliably achievable. However, under circumstances where space or weight are overriding concerns, the thickness of the primary barrier may be tapered due to the oblique path of the radiation, which increases both the distance to the barrier and also the effective (slant) thickness through the barrier. Likewise, oblique angles might be associated with small use factors.

When the radiation is obliquely incident on a barrier, the required thickness of the barrier will be less than that obtained by the above calculations. The difference between these thicknesses depends on: (1) the angle of obliquity (θ) between the radiation direction and the normal to the barrier, (2) the barrier material, (3) the required attenuation, and (4) the energy of the radiation. If there were no radiation scattering in the barrier material, the relationship between the computed slant thickness (t_S) and the actual thickness (t) of a barrier for obliquely incident radiation is given by $t/\cos\theta$, as illustrated in Figure 2.3. However, for large angles of obliquity, scattered photons may have a path length $<t_S$ before emerging from the barrier. This effect may necessitate a thickness of barrier $>t$. For most practical situations the effect is small and can be treated as a small increase in the approximate thickness t.

However, if the required attenuation is orders of magnitude, and the angle of obliquity is large (>45 degrees), the increase for concrete barriers is ~2 *HVL* for low-energy photons and ~1 *HVL* for high-energy photons.

The above approximate determinations are for radiation incident at a single angle. If the beam is very divergent, the angle of obliquity should not be used for the central ray because some of the

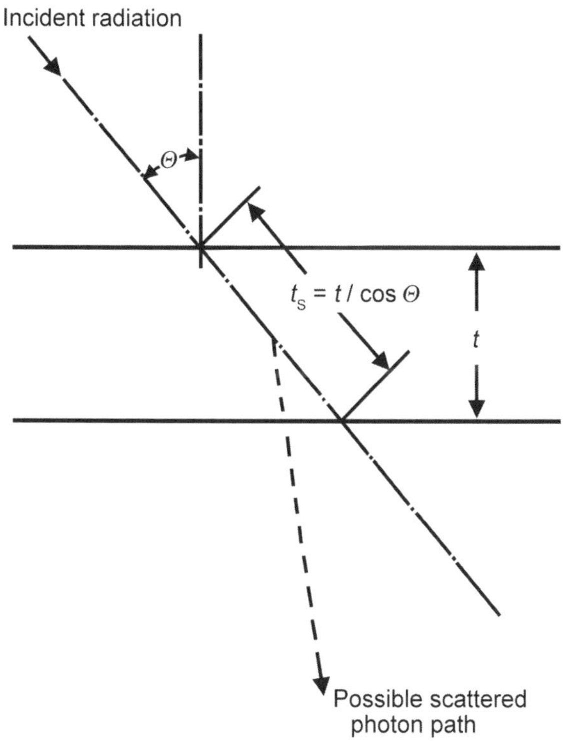

Fig. 2.3. Relationship between the slant thickness ($t_S = t/\cos\,\theta$) of radiation incident on a barrier with angle of obliquity (θ) and thickness of the barrier (t). Also shown is a scattered photon with a path length $<t_S$.

radiation will have a somewhat smaller angle of obliquity. Use of the minimum angle of obliquity will provide more attenuation than is required. Thus, judgment must be used in selecting the proper angle for divergent beams (for more details see Biggs, 1996; Kirn *et al.*, 1954). Also, by similar reasoning, the obliquity is usually taken into consideration only for primary radiation beams since the leakage and scattered-radiation sources can be too diffuse to apply a specific angle of incidence.

Laser lights used to align the patient in the primary beam may be recessed in the concrete. This recess thickness can be equivalent to about an *HVL* for high-energy radiation, so a steel or lead plate with a thickness providing the same attenuation as the removed concrete *should* be used behind the laser. Since the lasers generally require a mounting plate to allow for lateral adjustments in the position of the laser unit, the two functions can be combined in one plate.

2.2.2 *Barrier Widths*

As a general rule, the barrier width for the primary beam is determined by calculating the size of the diagonal of the largest beam and adding at least 30 cm to each side. If the primary barrier protrudes into the room, the maximum size of the beam is calculated in the plane of the inner part of the secondary barrier (*i.e.*, the target side) (Figure 2.4a). If the primary barrier protrudes outside the room, the maximum size of the beam is calculated in the plane of the outer part of the primary barrier (Figure 2.4b). If a composite primary barrier is constructed with concrete and either lead or steel, the size of the beam is calculated from the surface of the lead or steel distal to the target (Figure 2.4c). However, as shown by Taylor *et al.* (1999), for scattered radiation at 20 degrees or less (since scatter fractions increase rapidly with accelerating voltage and scattered-beam energies approach the primary-beam energy), the 30 cm margin may not be adequate for the higher primary-beam energy if the barrier does not intercept at least the 20 degree scattered radiation.

On most linear accelerators, although the largest field size is (40 × 40) cm^2 at 100 cm source-to-surface distance, the maximum size of the primary field is limited to ~50 cm diameter at 100 cm source-to-surface distance, equivalent to a half angle of 14 degrees. The barrier width is determined at the top of the primary wall barrier that is furthest from the isocenter and this width is maintained constant over the primary-barrier region (*i.e.*, both sidewalls and ceiling). Note that for a wide room with a low ceiling height, the width of the primary beam directly overhead may be considerably narrower than this barrier width. However, this allows for ease of construction since, otherwise, a more complicated form arrangement would be required to provide a tapered primary barrier. Alternatively, part of the ceiling primary barrier can include either lead or steel.

These materials come conveniently in either sheets (steel) or bricks (lead), so that the high-density shielding can easily be laid over the primary area to include the tapering of the beam across the surface of the ceiling. In any case, although the required thickness of a barrier may decrease as the beam moves to more oblique angles, the required width of the barrier will increase since the projected beam widens with distance.

Most radiation therapy facilities are designed with the gantry rotation plane orthogonal to the primary barriers. There are occasions, however, when, for reasons of space, esthetics or patient set-up convenience, it is desired to place the linear accelerator in the room with its axis of gantry rotation at, for example, 45 degrees with respect to the walls of the room. This geometry is sketched in Figure 2.4d. Great care has to be exercised in the design of the primary barrier for this situation, since photons traveling along the two opposite diagonal edges of the beam traverse the shielding at different angles. Thus, the position along the outside of the shielded wall at which the two edges of the primary-beam strike the barrier can be quite asymmetric with respect to the central axis of the beam. These two positions are denoted by A and B in the diagram. This effect depends on the thickness of the shielding and is greatest for concrete-only shields; conversely, if either lead or steel is used as part of the primary barrier, this effect is moderated since the barrier thickness may be considerably reduced.

2.2.3 *Laminated Barriers*

As stated above, there are situations when the primary barrier is not composed solely of homogeneous ordinary concrete (density 2.35 g cm^{-3}) (see Glossary). This is often the case when space constraints are paramount, and ordinary concrete is used in conjunction with steel or lead. For the primary photon beam, the total transmission factor is the product of the transmission factors of each of the individual materials in the barrier (*e.g.*, $B_T = B_{conc}\, B_{Pb}\, B_{steel}$, for concrete, lead and steel, respectively). However, this does not take into account the attenuation and production of photoneutrons and neutron capture gamma rays that must be considered if the primary-beam accelerating voltage is above 10 MV. In such high-energy cases, if a composite barrier design (*e.g.*, steel or lead plus concrete) is not carried out correctly, the metal layer can become a photoneutron source potentially resulting in an increased exposure problem beyond the shield. This was first noted by McGinley *et al.* (1988), with further data in McGinley (1992a; 1992b). It is worth noting that this is a problem only for primary barriers and not for secondary barriers, since scattered radiation well beyond the primary barrier is not energetic enough to produce photoneutrons, and the leakage-radiation intensity, when combined with cross sections for photoneutron production, does not produce a significant number of neutrons in secondary barriers. The problem of calculating laminated shielding was addressed by McGinley (1992a; 1992b) and by McCall and Kleck (1994). Since the McGinley (1992a; 1992b) method is very straightforward and covers most situations, it is discussed below.

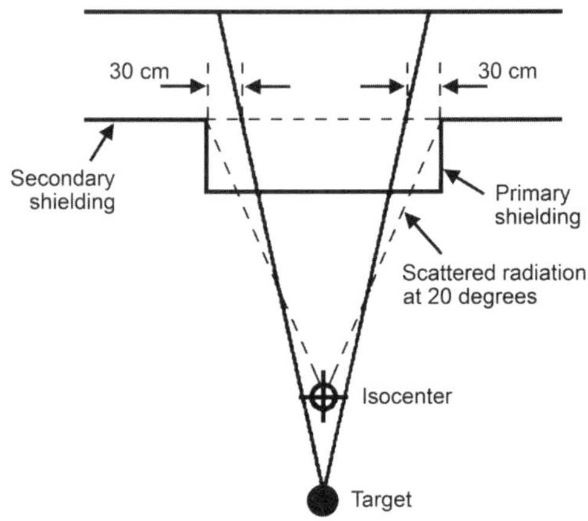

Fig. 2.4a. Width of primary barrier protruding into the room.

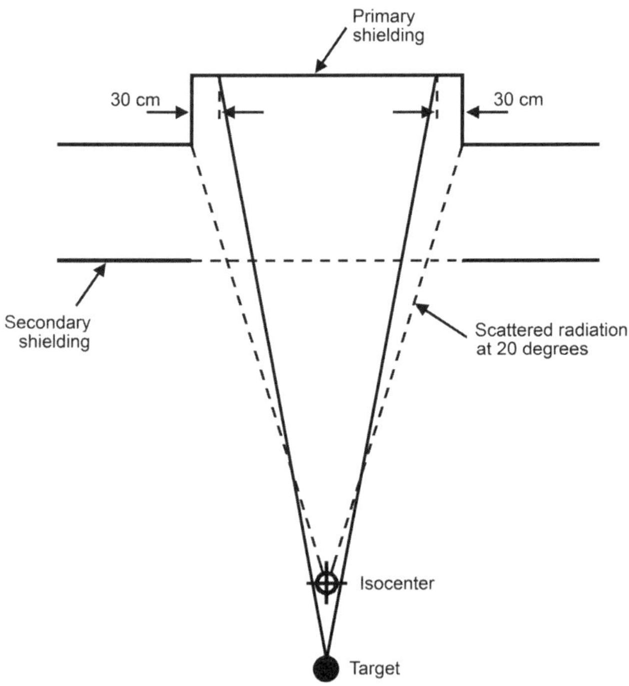

Fig. 2.4b. Arrangement for the primary barrier when the inside wall is continuous.

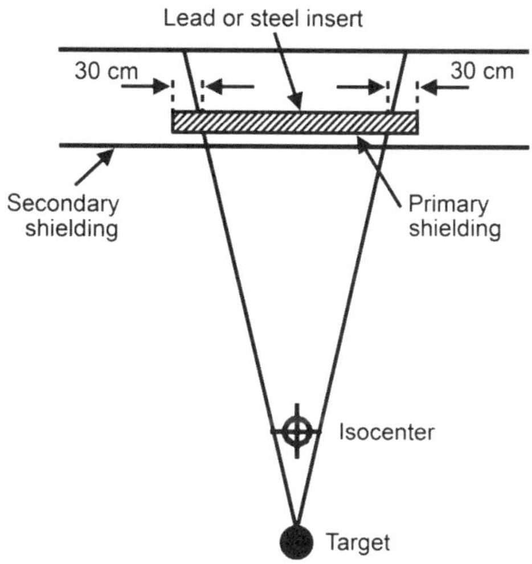

Fig. 2.4c. Arrangement for the primary barrier when lead or steel is used to maintain a uniform wall thickness.

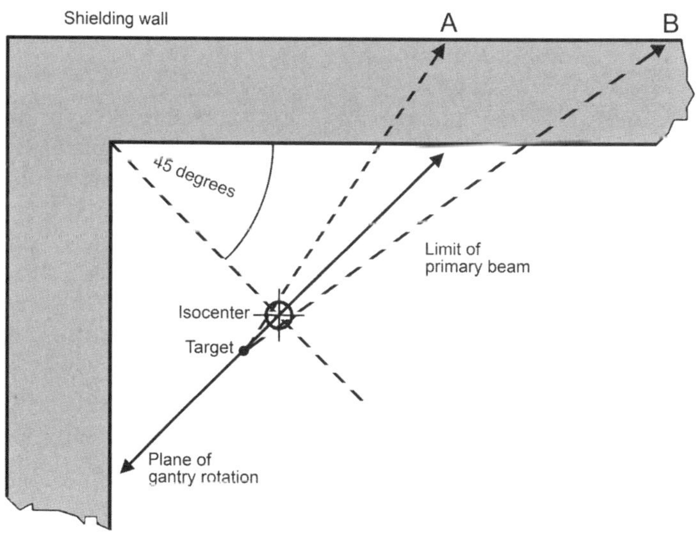

Fig. 2.4d. Sketch showing angulation of the plane of gantry rotation at 45 degrees to the walls. Note the asymmetry of the extremities of the primary beam on the outside of the wall (A, B) compared with the central axis of the beam.

The following empirical Equation 2.5 (McGinley, 1992a)[7] was used to estimate the neutron dose-equivalent per week beyond the laminated barrier when the collimator is opened to maximum size (Figure 2.5).

$$H_n = \frac{D_o\, R\, F_{max}}{\left(\frac{t_m}{2} + t_2 + 0.3\right)} \left[10^{-\left(\frac{t_1}{TVL_x}\right)}\right]\left[10^{-\left(\frac{t_2}{TVL_n}\right)}\right] \tag{2.5}$$

In Equation 2.5:

H_n = neutron dose equivalent per week (μSv week^{-1})
D_o = x-ray absorbed dose per week at isocenter (cGy week^{-1})
R = neutron production coefficient (in neutron microsievert per x-ray centigray per beam area in m^2) (*i.e.*, μSv cGy^{-1} m^{-2})
F_{max} = maximum field area at isocenter (m^2)
t_m = metal slab thickness (meters)
t_1 = first concrete slab thickness (meters)
t_2 = second concrete slab thickness (meters)
TVL_x = tenth-value layer in concrete for the primary x-ray beam (meters) (Table B.2)
TVL_n = tenth-value layer in concrete for neutrons (meters) (can be extracted from Figure A.2)
0.3 = distance from the outer surface of the barrier to the point of occupancy (meters)

McGinley (1992a) has reported on accelerators operated at 18 MV and measured neutron production coefficients (R) of 19 and 1.7 μSv cGy^{-1} m^{-2} for lead and steel, respectively; while R is decreased to around 3.5 μSv cGy^{-1} m^{-2} for lead at 15 MV.

For the low-energy spectrum of neutrons produced by medical accelerators, Kase *et al.* (2003) measured a TVL_n of 45 g cm^{-2} in ordinary concrete. The dose equivalent from neutron capture gamma rays is implicitly taken into account in these measurements. Hence a value of 25 cm would be a conservatively safe estimate[8] of the TVL_n for ordinary concrete as well as heavy concretes since the hydrogen content does not vary significantly among them (Table B.3).

[7]In order for the units in empirical Equation 2.5 to be the same for each side of the equation, it is understood that the denominator ($t_m/2 + t_2 + 0.3$) is divided by a unit meter, so that the denominator is unitless.

[8]This value is based on the TVL_n = 45 g cm^{-2} and a density of 2.3 g cm^{-3} for a type of ordinary concrete: 45 g cm^{-2}/2.3 g cm^{-3} = 19.6 cm; then increased to 25 cm as a conservatively safe estimate.

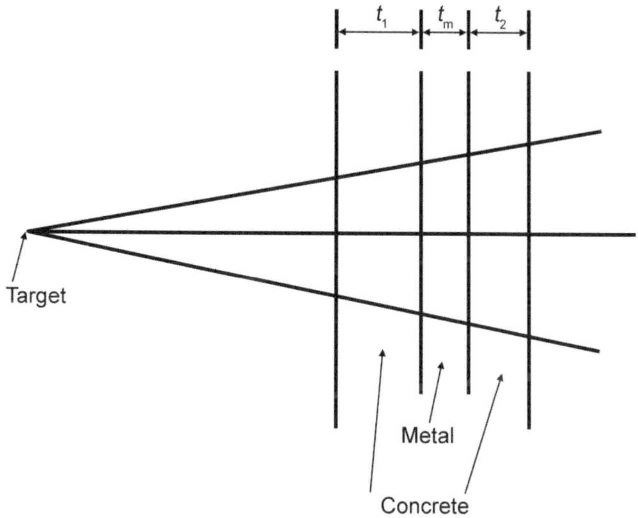

Fig. 2.5. Laminated barrier with metal of thickness t_m between concrete thicknesses of t_1 and t_2.

In the process of neutron production in lead or steel and subsequent neutron interactions in the primary concrete (or hydrogenous) barrier, gamma rays are produced. Some of these are neutron capture gamma rays in the concrete while others are emitted from the lead or steel nuclei after they undergo the photoneutron interaction and are left in an excited state from which they subsequently emit de-excitation gamma rays. The energies and intensities of these two components are not well characterized at this time, but they are certainly geometry and material dependent. McGinley and Butker (1994) have examined measurements from several rooms with laminated ceilings. Based on their measurements with steel and concrete laminates at 15 and 18 MV photon beam energies, they conclude that, if the calculated transmitted x-ray dose-equivalent component (H_{tr}) is multiplied by 2.7, it will yield a conservatively safe estimate of the photon dose equivalent (from x plus gamma rays) in sievert (H_{phtr}). Thus the total dose equivalent beyond the barrier (H_{Tot}) is:

$$H_{Tot} = H_n + H_{phtr} = H_n + 2.7\ H_{tr} \tag{2.6}$$

When B_{pri} is known, the value of H_{tr} may be obtained from Equation 2.1 with P replaced by H_{tr}. If the sum H_{Tot} is $>P$, then the calculation is iterated to reduce H_{tr} further until H_{Tot} achieves the shielding design goal.

2.3 Secondary Barriers

Secondary barriers need to be designed to adequately protect individuals beyond the accelerator room from: (1) leakage radiation, (2) scattered radiation from the patient, (3) scattered radiation from the walls, and (4) secondary radiations (including photoneutrons and neutron capture gamma rays) produced in the accelerator head or in scattering throughout the room. When dealing with secondary barriers, photoneutrons and neutron capture gamma rays are usually a concern only for photon energies above 10 MeV and when dealing with thin barriers such as the doors in a maze or HVAC conduits. These are considered in Section 2.4.2 on maze and door design.

Since leakage radiation and scattered radiation are of such different energies, the secondary-barrier requirements of each are typically computed separately and compared in order to arrive at the final recommended thickness.

The barrier transmission needed for radiation scattered by the patient (B_{ps}) is given by Equation 2.7.

$$B_{ps} = \frac{P}{aWT} d_{sca}^2 \, d_{sec}^2 \, \frac{400}{F} \tag{2.7}$$

In Equation 2.7, the symbols *P*, *W* and *T* are as defined earlier (Section 2.2.1) and:

- d_{sca} = distance from the x-ray target to the patient or scattering surface (meters)
- d_{sec} = distance from the scattering object to the point protected (meters)
- a = scatter fraction or fraction of the primary-beam absorbed dose that scatters from the patient at a particular angle (see Table B.4 in Appendix B)
- F = field area at mid-depth of the patient at 1 m (cm^2)

and the value 400 assumes the scatter fractions are normalized to those measured for a 20 cm × 20 cm field size. The distances d_{sca} and d_{sec}, and the Area *F* are shown in Figure 2.6.

Note that the use factor for patient-scattered radiation is taken as one in Equation 2.7. Strictly speaking, *U* is a function of the gantry angle. However, if the calculation is performed with the minimum angle of scatter from the patient to the point of calculation and a use factor of one is also used, the barrier thickness will be overestimated due to the conservatively higher scatter fraction from the smaller scattering angles.

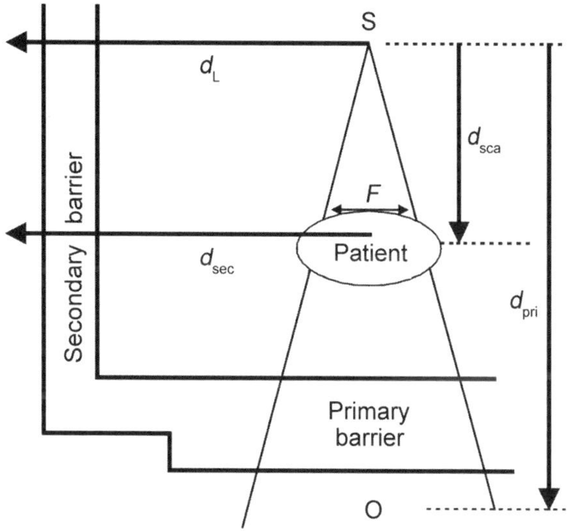

Fig. 2.6. Room layout showing distances associated with patient-scattered (d_{sca}, d_{sec}) and leakage radiations (d_L).

As noted, the scattered-radiation energy is significantly degraded (beyond 20 degree scattered radiation) from that of the primary beam and thus separate data are used to compute its transmission through the barrier. Tables B.5a and B.5b give *TVL* values in concrete and lead, respectively, for radiations scattered from the patient at different scattering angles and beam energies. For other materials, the *TVL* for the patient-scattered radiation can be estimated by using the mean energy of the scattered radiation from Table B.6 (Appendix B) and the *TVL* values from Figures A.1a and A.1b (Appendix A).

The barrier transmission of leakage radiation alone (B_L) is given by Equation 2.8.

$$B_L = \frac{P\, d_L^2}{10^{-3}\, W\, T} \tag{2.8}$$

In Equation 2.8, the factor 10^{-3} arises from the assumption that leakage radiation from the accelerator head is 0.1 % of the useful beam. The use factor again is taken as one, and d_L is measured from the isocenter if it can be assumed that the accelerator gantry angles used are, on average, symmetric. If this is not the situation, then the distance to the individual barriers *should* be taken from the closest approach of the accelerator head to each barrier and actual use factors *should* be employed in the denominator of Equation 2.8. Table B.7 can be used to find measured *TVLs* for leakage radiation for ordinary concrete. If the clinical practice includes IMRT, then the workload for leakage radiation (W_L) *shall* be modified in accordance with Section 3.2.2. In non-IMRT situations, W_L is equal to *W*.

After the secondary-barrier transmission factor is determined for both leakage and scattered radiation, the required thickness of the shielding material for each contribution can be determined by the use of tenth-value layers (Tables B.5a, B.5b, and B.7 in Appendix B, or Figures A.1a and A.1b in Appendix A) along with Equations 2.2 and 2.3. If the thickness of the required barrier is about the same for each secondary component (*i.e.*, as though the occupied space in question is irradiated by two sources of approximately equal intensity), 1 *HVL* is added to the larger of the two barrier thicknesses. If the two thicknesses differ by a *TVL* or more, the larger barrier thickness is used. This is often referred to as the two-source rule. In most high-energy accelerator facilities, a secondary barrier that is adequately designed for the leakage-radiation component will be more than adequate for the scattered radiation with the possible exception of zones adjacent to the primary barrier intercepted by small angle scatter.

When Equations 2.1, 2.7, and 2.8 are solved for *P*, they yield the dose equivalent of the primary, scattered and leakage radiations, respectively. A quality factor of unity is assigned for the conversion of absorbed doses (as represented by the workload *W*) to dose equivalents for low-LET radiation components that are transmitted to and measured at the shielded area.

2.4 Doors and Mazes

Entryways to accelerator rooms that use a maze design present some unique considerations as secondary barriers. This results from the scattering properties of the radiations and the desire to keep the doors in such designs as light as possible. The maze design is treated under two separate headings: low-energy accelerators ($\leq$10 MV) and high-energy accelerators (>10 MV) since there are major differences in the secondary radiation types and fluences produced in each of these cases.

2.4.1 *Low-Energy Accelerators*

A maze such as the one shown in Figure 2.7 is commonly used to reduce the radiation level at the entrance to the accelerator room so a massive door is not required. However, unless the maze is very long or has multiple legs the door shielding must still be evaluated. In the method described here, the dose equivalent[9] at the position of the maze door is evaluated first for the case in which the beam is directed perpendicular to Wall G of Figure 2.7. Next, a simple empirical equation is used, that relates the dose equivalent determined in the first step to the total dose equivalent at the maze door produced by beams aimed in the major beam directions (up, down, left and right) at the maze door. Finally, the thickness of shielding material required to reduce the dose equivalent to the shielding design goal (or less) is evaluated.

The radiation reaching the maze door is due to scattering of photons from the room surfaces and patient as well as direct penetration of head-leakage radiation through the inner maze Wall Z. These components are given as follows:

H_S = dose equivalent per week due to scatter of the primary beam from the room surfaces

H_{LS} = dose equivalent per week due to head-leakage photons scattered by the room surfaces

H_{ps} = dose equivalent per week due to primary beam scattered from the patient

H_{LT} = dose equivalent per week due to leakage radiation which is transmitted through the inner maze wall

Equation 2.9 is used to determine the radiation scattered to the maze door when the primary beam strikes Wall G. This is a modification of the treatment of this situation as originally given in NCRP (1977) and later modified by Numark and Kase (1985).

$$H_S = \frac{W\, U_G\, \alpha_0\, A_0\, \alpha_z\, A_z}{(d_h\, d_r\, d_z)^2} \tag{2.9}$$

In Equation 2.9:

H_S = dose equivalent per week at the maze door due to scattering of the primary beam from Wall G

W = workload (Gy $week^{-1}$)

[9]In this Report, the dose equivalent (H) (sievert) for photons is equivalent to the absorbed dose (D) (gray) at a given location, since the quality factor for photons is assigned a value of unity (NCRP, 1993)

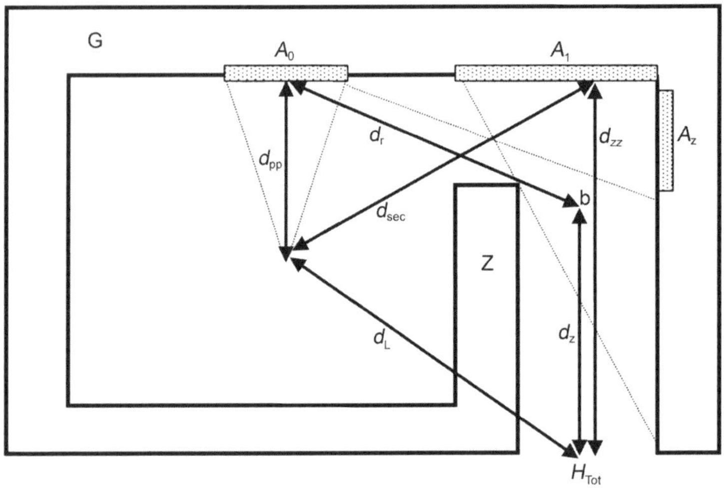

Fig. 2.7. General room layout for definition of parameters used in maze door shielding (see Figure 7.1 for more detail).

U_G = use factor for the Wall G

α_0 = reflection coefficient at the first scattering surface A_0

A_0 = beam area at the first scattering surface (m^2)

α_Z = reflection coefficient for second reflection from the maze surface A_z (an energy of 0.5 MeV is usually assumed)

A_Z = cross-sectional area of maze inner entry projected onto the maze wall from the perspective of the irradiated primary barrier A_0 (m^2)

d_h = perpendicular distance from the target to the first reflection surface [equal to d_{pp} (perpendicular distance from isocenter to the wall, see Figure 2.7) plus 1 m] (meters)

d_r = distance from beam center at the first reflection, past the edge of the inner maze wall, to Point b on the midline of the maze (meters)

d_z = centerline distance along the maze from Point b to the maze door (meters)

Values of the reflection coefficients (α) for normal and 45 degree incidence on concrete, lead and steel have been estimated using Monte-Carlo methods (IAEA, 1979; Lo (1992). The reported values, which agree within factors of two to three, have been evaluated by NCRP and suggested values are given in Tables B.8a through B.8f (Appendix B).

McGinley (2002) reports that this calculation should be restricted to facilities at which the height-to-width ratio of the maze is between one and two, and agrees with NCRP (1977) that the value of [d_z /(maze height × maze width)$^{1/2}$] should be between two and six. Though the second condition may not be met, since many facilities are designed with relatively short mazes, it was still found that agreement was within a factor of two for most cases. The height-to-width ratio can often be achieved by employing a simple lintel over the inner maze entrance.

Head-leakage radiation can strike Wall G over Area A_1 and undergo a single scatter before reaching the maze door. Equation 2.10 is used to evaluate this dose-equivalent component at the door (McGinley and James, 1997).

$$H_{LS} = \frac{L_f \, W_L \, U_G \, \alpha_1 \, A_1}{(d_{sec} \, d_{zz})^2} \tag{2.10}$$

In Equation 2.10:

H_{LS} = dose equivalent per week at maze door due to single-scattered head-leakage radiation

L_f = head-leakage radiation ratio at 1 m from the target [taken as 1/1,000 or 0.1 % per the IEC (2002) requirement]

W_L = workload for leakage radiation (Gy week^{-1}) (which can be different than the primary workload as discussed in Section 3.2.2)

U_G = use factor for the Wall G

α_1 = reflection coefficient for scatter of leakage radiation from Wall G

A_1 = area of Wall G that can be seen from the maze door (m^2)

d_{sec} = distance from the target to the maze centerline at Wall G (meters) [note this may be measured from the isocenter as representing the average target position]

d_{zz} = centerline distance along the maze (meters)

The publication by Nelson and LaRiviere (1984) bases the reflection coefficient α_1 in Equation 2.10 on an effective energy of 1.4 MeV for 6 MV x rays and 1.5 MeV for 10 MV x rays. Values for such modal bremsstrahlung energies may be obtained from Table B.8a in Appendix B. Higher energies will be dealt with in Section 2.4.2 since they entail neutron capture processes.

Patient-scattered radiation to the maze door is calculated by use of Equation 2.11 (McGinley and James, 1997). Figures 2.6 and 2.7 depict the distances and areas used to calculate the dose equivalent due to patient-scattered radiation.

$$H_{ps} = \frac{a(\theta)\, W\, U_G \left(\frac{F}{400}\right) \alpha_1\, A_1}{(d_{sca}\, d_{sec}\, d_{zz})^2} \qquad (2.11)$$

In Equation 2.11:

H_{ps} = dose equivalent per week at the maze door due to patient-scattered radiation

$a(\theta)$ = scatter fraction for patient-scattered radiation at angle θ (from Table B.4 in Appendix B)

W = workload for the primary beam (Gy week^{-1})

U_G = use factor for the Wall G

F = field area at mid-depth of the patient at 1 m (cm^2)

α_1 = reflection coefficient for Wall G for the patient-scattered radiation

A_1 = area of Wall G that can be seen from the outer maze entrance (m^2)

d_{sca} = distance from the target to the patient (meters)

d_{sec} = distance from the patient to Wall G at the maze centerline (meters)

d_{zz} = centerline distance along the maze length from the scattering surface A_1 to the door (meters)

The reflection coefficient α_1 may be obtained for the average energy of the photons scattered at various angles by the patient, but a conservatively safe calculation results if an energy of 0.5 MeV is used when determining this coefficient. When the endpoint energy of the accelerator is >10 MV, patient-scattered radiation is usually ignored since it becomes insignificant in comparison to the dose equivalent produced by leakage radiation and the neutron capture gamma rays produced in the maze by slow neutrons (*i.e.*, neutrons with kinetic energies of ~1 eV to a few kiloelectron volts).

The leakage radiation that is transmitted through maze Wall Z to the treatment room door is estimated by use of Equation 2.12.

$$H_{LT} = \frac{L_f\, W_L\, U_G\, B}{d_L^2} \qquad (2.12)$$

In Equation 2.12:

H_{LT} = dose equivalent per week at the maze door due to leakage radiation which is transmitted through the inner maze wall

L_f = head-leakage radiation ratio, which is taken as a conservatively safe value of 10^{-3} of the useful beam

W_L = workload for leakage radiation (Gy week^{-1}) (which can be different than the primary workload as discussed in Section 3.2.2)

U_G = use factor for the gantry orientation G

B = transmission factor for Wall Z along the oblique path traced by d_L

d_L = distance from the target to the center of the maze door through the inner maze wall (meters).

After each of the individual components has been calculated, the total dose equivalent (H_G) at the maze door, with the beam aimed at Wall G (Figure 2.7), is obtained by summing the dose-equivalent components considered above.

$$H_G = f\,H_S + H_{LS} + H_{ps} + H_{LT} \tag{2.13}$$

Note that the workload with the beam aimed at Wall G (*i.e.*, $W\,U_G$) is employed for the calculation of H_G, and the fraction of the primary beam transmitted through the patient is represented by f in the equation. For example, f has a value of ~0.25 for 6 to 10 MV x rays when the field size is (40 × 40) cm^2 and a (40 × 40 × 40) cm^3 phantom is utilized (McGinley and James, 1997).

Finally, when the use factors for the major beam directions (0, 90, 180 and 270 degrees) are each taken as one-quarter, the total dose equivalent (H_{Tot}) at the maze door from photon leakage radiation and scattered radiation is not simply 4 H_G, but is estimated as 2.64 H_G (McGinley, 2002), where a quality factor of unity is assigned for the photons from low-energy accelerators (≤10 MV).

$$H_{Tot} = 2.64\,H_G \tag{2.14}$$

Equation 2.14 is to be used with caution if the room design is different than that shown in Figure 2.7 [*i.e.*, $2 < d_{zz} / \sqrt{(\text{maze width} \times \text{height})} < 6$; and $1 < (\text{maze height/maze width}) < 2$].

The transmission factor required for the door shielding is calculated by dividing the shielding design goal (*P*) needed for the area outside the door by H_{Tot}. For accelerators with energies at or below 10 MV, the door shielding is based on broad-beam transmission data for 0.2 MeV photons (Al-Affan, 2000; McGinley and James, 1997). If the inner maze wall is very thin, then the spectrum and intensity of the radiation at the door will increase due to the increased leakage radiation transmitted through this wall. In that case, the shielding requirement will be determined by H_{LT} alone and the transmission data for the leakage-radiation energy. In general, this will not be the case if H_{LT} is less than half of H_G.

The above technique also assumes a nearly uniform distribution of gantry use factors around the treatment rotational plane and if this is not the case (*e.g.*, for TBI procedures) the empirical factor of 2.64 may not be valid.

When the endpoint energy of the accelerator is above 10 MV, the techniques outlined above are still valid. However, the presence of photoneutrons and neutron capture gamma rays must also be considered as discussed in Section 2.4.2.

2.4.2 *High-Energy Accelerators*

The estimate of the dose equivalent from photons scattered through the maze can be made using the method given above, however since the average energy of neutron capture gamma rays from concrete is 3.6 MeV (Tochilin and LaRiviere, 1979), a maze and door that provide sufficient shielding for the neutron capture gamma rays will also be adequate for the scattered photons.

For mazes in high-energy accelerator rooms, where the distance from A to B in Figure 2.8 is >2.5 m, the photon field is dominated by neutron capture gamma rays and the scattered photon component can be ignored. In fact, the photon dose equivalent outside the maze door changes only slightly when the collimator of the accelerator is adjusted from maximum size to the closed position or when the scattering phantom is removed from the beam (McGinley and Huffman, 2000). Therefore door shielding in high-energy rooms is usually dominated by the neutron capture gamma ray and photoneutron requirements.

2.4.2.1 *Photon Dose-Equivalent Calculation at the Maze Door.* A method for estimating the neutron capture gamma-ray dose equivalent at the treatment room door has been described by McGinley *et al.* (1995). The dose equivalent (h_φ) from the neutron capture gamma rays at the outside maze entrance, per unit absorbed dose of x rays at the isocenter, is given by Equation 2.15.

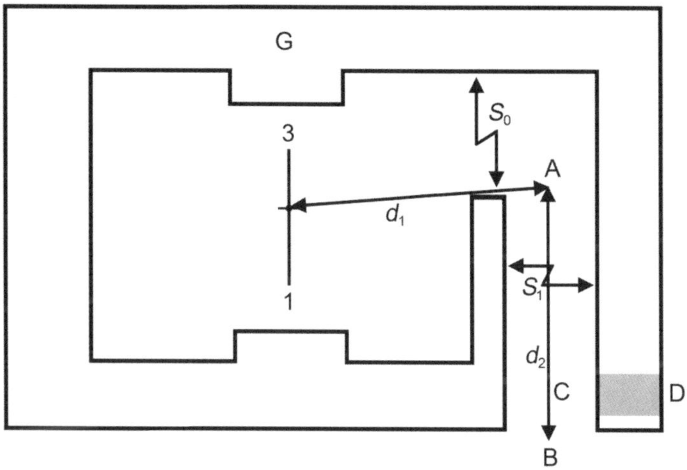

Fig. 2.8. Room layout for calculating neutron capture gamma-ray and neutron dose equivalents at the maze door.

$$h_{\varphi} = K\ \varphi_{\mathrm{A}}\ 10^{-\left(\frac{d_2}{TVD}\right)} \tag{2.15}$$

In Equation 2.15:

K = ratio of the neutron capture gamma-ray dose equivalent (sievert) to the total neutron fluence at Location A in Figure 2.8 (an average value of 6.9×10^{-16} Sv m^2 per unit neutron fluence was found for K based on measurements carried out at 22 accelerator facilities)[10]

φ_{A} = total neutron fluence (m^{-2}) at Location A per unit absorbed dose (gray) of x rays at the isocenter

d_2 = distance from Location A to the door (meters)

TVD = tenth-value distance[11] having a value of ~5.4 m for x-ray beams in the range of 18 to 25 MV, and a value of ~3.9 m for 15 MV x-ray beams

[10]McGinley, P.H. (1998). Personal communication (Emory University, Atlanta, Georgia).

[11]This technique (Sections 2.4.2.2.1 and 2.4.2.2.2) employs the concept of a tenth-value distance (*TVD*), which is the distance required for the photon fluence to decrease 10-fold.

The total neutron fluence at the inside maze entrance (Location A in Figure 2.8) per unit absorbed dose from x rays at the isocenter can be evaluated by use of Equation 2.16 (McCall *et al.*, 1999; NCRP, 1984).

$$\varphi_A = \frac{\beta Q_n}{4\pi d_1^2} + \frac{5.4\ \beta Q_n}{2\pi S_r} + \frac{1.3\ Q_n}{2\pi S_r} \tag{2.16}$$

In Equation 2.16, the three terms represent the direct, scattered and thermal neutron components, respectively; and:

β = transmission factor for neutrons that penetrate the head shielding (one for lead and 0.85 for tungsten head shielding)

d_1 = distance from the isocenter to Location A in Figure 2.8 (meters)

Q_n = neutron source strength in neutrons emitted from the accelerator head per gray of x-ray absorbed dose at the isocenter

S_r = total surface area of the treatment room (m^2)

and $1/(2\pi)$ in the scattered and thermal neutron terms accounts for the fraction of the neutrons that enter the maze

Neutron source strength (Q_n) values for some typical accelerators and nominal energies are given in Table B.9 (Appendix B), as reported by Followill *et al.* (2003) and McGinley (2002). A plot of these data demonstrates the variability of the measured data for a given stated energy as well as the rapid increase in photoneutron production that occurs as the beam accelerating voltage rises above 10 MV. Though photoneutrons have been measured in both the treatment room and at the maze entrance of 10 MV accelerators, their fluence is low enough that they do not typically pose a radiological hazard (Deye and Young, 1977). There have, however, been reports of photoneutron production in lead radiation barriers with nominal 10 MV beams (McGinley and Butker, 1994), but on closer inspection this beam had an effective accelerating voltage of 11.6 MV based on an American Association of Physicists in Medicine (AAPM) Task Group 21 endpoint energy measurement.[12] Thus it is very important to ensure that a "nominal" 10 MV beam is not in fact closer to a 12 MV beam due to the tuning of the accelerator.

[12]McGinley, P.H. (1994). Personal communication (Emory University, Atlanta, Georgia).

The weekly dose equivalent at the door due to neutron capture gamma rays (H_{cg}), in sievert per week, is then the product of the workload for leakage radiation (W_L) and h_φ:

$$H_{cg} = W_L \, h_\varphi \tag{2.17}$$

2.4.2.2 *Neutron Dose Equivalent at the Maze Door.* Medical accelerators operating above 10 MV require door shielding for neutrons as well as photons. The neutron dose equivalent at the maze entrance varies somewhat, but is largely unaffected by the collimator setting (McGinley and Huffman, 2000). Since the maximum neutron fluence is obtained by closing the collimators, it is expected that most photoneutrons originate in the head of treatment accelerators (Kase *et al.*, 1998; Mao *et al.*, 1997). The neutron field in the maze is also a function of the gantry angle and location of the target rotational plane in the treatment room. For example, McGinley and Butker (1991) reported that the neutron level at the treatment room door was maximum when the gantry angle was aligned along the horizontal line marked 3–1 in Figure 2.8. This alignment places the neutron source (the target) closer to the inner maze entrance than other positions of the gantry and the target in the rotational plane. For this room layout, the neutron dose equivalent at the maze door was found to vary by a factor of two as the gantry angle was changed.

Determination of the neutron dose equivalent at the outer maze entrance can be carried out using one of several analytical techniques outlined below. These techniques all employ the concept introduced earlier by Maerker and Muckenthaler (1967) of a tenth-value distance (TVD) for the fall off of thermal neutron fluences through mazes and large ducts. Their data were represented in Figure F.11 in NCRP (1977) and indicate that the TVD is roughly equal to three times the square root of the product of the height times the width of the opening and that each additional leg in the opening decreases the fluence roughly another three-fold.

2.4.2.2.1 *Kersey's method.* One of the earliest techniques for evaluating the neutron fluence at a maze entry was given by Kersey (1979). In this formulation, the effective neutron source position is taken to be the isocenter of the accelerator, and the neutron dose equivalent ($H_{n,D}$) at the outside maze entrance per unit absorbed dose of x rays at the isocenter is given by Equation 2.18.

$$H_{n,D} = (H_0)\left(\frac{S_0}{S_1}\right)\left(\frac{d_0}{d_1}\right)^2 10^{-\left(\frac{d_2}{5}\right)} \tag{2.18}$$

In this application of Kersey's method, H_0 is the total (direct plus room-scattered plus thermal) neutron dose equivalent at a distance d_0 (1.41 m) from the target per unit absorbed dose of x rays at the isocenter (mSv Gy^{-1}) (see Table B.9 in Appendix B for measured values of H_0). The ratio S_0/S_1 is the ratio of the inner maze entrance cross-sectional area to the cross-sectional area along the maze (Figure 2.8). These are usually different primarily because of their different widths, though the height may change also due to the use of lintels above the inner maze entrance. Distance d_1, which is shown in Figure 2.8, is the distance from the isocenter to the point on the maze centerline from which the isocenter is just visible (A). For a maze with one bend as illustrated, d_2 is the distance in meters from A to B. In the case of a maze with two bends, d_2 is the distance from A to C plus length C to D. Note that for this method the maze has a *TVD* of 5 m for the attenuation of neutrons in the maze.

Kersey's technique has been evaluated by McGinley and Butker (1991) for a number of treatment facilities and for a variety of accelerator models produced by several manufacturers. The accelerating voltage of the accelerators investigated ranged from 15 to 18 MV. It was determined that the ratio of the neutron dose equivalent calculated by use of Kersey's method divided by the measured neutron dose equivalent varied from 0.82 to 2.3 for the 13 accelerator facilities investigated in the study. McGinley and Butker (1991) found that the *TVD* for maze neutrons was ~16 % < 5 m. Therefore, 5 m is a conservatively safe value to use for the neutron *TVD* when the dose equivalent is determined at the maze outer entrance. It was also discovered that a second turn in the maze reduced the neutron level by a factor of at least three as compared to the value obtained by Kersey's equation. Therefore they suggested modifications to Kersey's method, which results in a more realistic approach.

2.4.2.2.2 *Modified Kersey's method.* Because of these findings, McGinley and Huffman (2000) modified Kersey's equation to obtain a better representation of the neutron dose equivalent along the maze. It was found that the neutron dose equivalent per unit absorbed dose of x rays at the isocenter (Sv Gy^{-1}) as a function of the distance d_2 along the maze centerline could be resolved into the sum of two exponential functions.

Subsequently, Wu and McGinley (2003) pointed out that the approach in McGinley and Huffman (2000) did not account for rooms with nonstandard surface areas or mazes with exceptional width or length, and they further refined the analysis of the measured data to give Equation 2.19 for the neutron dose equivalent along the maze length.

$$H_{n,D} = 2.4 \times 10^{-15}\ \varphi_A \sqrt{\frac{S_0}{S_1}} \left[1.64 \times 10^{-\left(\frac{d_2}{1.9}\right)} + 10^{-\left(\frac{d_2}{TVD}\right)}\right] \quad (2.19)$$

In Equation 2.19:

$H_{n,D}$ = neutron dose equivalent at the maze entrance in sievert per unit absorbed dose of x rays (gray) at the isocenter and thus the constant has units of Sv n^{-1} m^2

φ_A = neutron fluence per unit absorbed dose of photons (m^{-2} Gy^{-1}) at the isocenter as given by Equation 2.16

S_0/S_1 = ratio of the inner maze entrance cross-sectional area to the cross-sectional area along the maze (Figure 2.8)

TVD = tenth-value distance (meters) that varies as the square root of the cross-sectional area along the maze S_1 (m^2), *i.e.*:

$$TVD = 2.06 \sqrt{S_1} \quad (2.20)$$

The weekly dose equivalent (sievert per week) (H_n) at the door due to neutrons is thus given by:

$$H_n = W_L\ H_{n,D} \quad (2.21)$$

and W_L is the weekly leakage-radiation workload.

2.4.2.3 *Total Dose Equivalent at the Maze Door.* The total weekly dose equivalent at the external maze entrance (H_W) is then the sum of all the components from the leakage and scattered radiations (Equation 2.14), the neutron capture gamma rays (Equation 2.17) and the neutrons (Equation 2.21):

$$H_W = H_{Tot} + H_{cg} + H_n \quad (2.22)$$

For most mazes, where energies above 10 MV are used, H_{Tot} is an order of magnitude smaller than the sum of H_{cg} and H_n, and is therefore negligible.

2.4.3 *Door Shielding*

The shielding of a maze door for scattered and leakage photons is described in Section 2.4.1. However, the scattered and leakage dose equivalents are generally relatively low compared with the other components (photoneutrons and neutron capture gamma rays), which become energetically possible at higher energies (McGinley *et al.*, 2000). The average energy of neutron capture gamma rays is 3.6 MeV (Tochilin and LaRiviere, 1979), and can range as high as 10 MeV (NCRP, 1984) for very short mazes. Thus a lead *TVL* of 6.1 cm may be required (NCRP, 1984).

The average neutron energy at the maze entrance is reported to be ~100 keV, with a *TVL* in polyethylene of 4.5 cm (NCRP, 1984). Borated polyethylene (BPE) (5 % by weight) is only a little less effective in fast neutron shielding, but is much more effective for thermal neutrons compared with polyethylene without boron. The *TVL* for BPE is 3.8 cm for 2 MeV neutrons, and 1.2 cm for thermal neutrons. For purposes of maze door shielding, a conservatively safe recommendation is that a *TVL* of 4.5 cm be used in calculating the BPE thickness requirement.

Many accelerator rooms with maze lengths on the order of 8 m or greater will require 0.6 to 1.2 cm of lead and 2 to 4 cm of BPE for shielding in the door. One suggested arrangement of the lead and the BPE is: lead, BPE, lead. The lead on the source side of the BPE is to reduce the energy of the neutrons by nonelastic scattering, and hence make the BPE more effective in neutron shielding. The lead on the outside of the BPE will serve to attenuate the neutron capture gamma rays from the BPE with energy of 478 keV. Often, the outside lead will not be necessary when the maze is long enough to attenuate the neutrons sufficiently before they encounter the door (McCall, 1997).

2.4.4 *Alternate Maze and Door Designs*

The procedure outlined above for the design of doors for typical mazes may result in a heavy and expensive door that must be equipped with a motorized opener. It has been suggested (NCRP, 1984) that by keeping the neutrons out of the maze the door shielding can be reduced in thickness or eliminated completely. McGinley and Miner (1995) investigated three techniques, which are outlined below, to prevent neutrons from leaving the treatment room and entering the maze:

1. reduce the opening at the inside maze entrance;
2. add a light-weight door containing a thermal neutron absorber (boron 9 % by weight) at the inside maze entrance; and
3. place a BPE (5 % boron) door at the inside maze entrance.

Table 2.1 indicates the measured dose-equivalent rate from a nominal 18 MV accelerator at the outside maze entrance when each of these techniques was utilized, as compared to a conventional maze. Each maze length (d_2) was 6.5 m and the accelerators were operated at an x-ray absorbed-dose rate of 6.67×10^{-2} Gy s^{-1} (4 Gy min^{-1}) at the isocenter.

In Technique 1, the inside maze opening was reduced in area to (1.22×2.13) m^2 using concrete 45.7 cm thick surrounding the opening. A 7 mm thick panel, containing 8.9 % by weight boron [Boraflex® (Brand Industrial Services, Inc., Park Ridge, Illinois)] was employed for Technique 2, and a 5 cm thick polyethylene (5 % boron) door was used for Technique 3. Technique 3 produced the largest reduction in total dose equivalent and only a relatively thin lead sheet, ~1 cm in thickness, was needed in the outer maze door. Lalonde (1997) and Uwamino *et al.* (1986) also reached similar conclusions based on measurements made at the outside maze entrance when the inside maze entrance was blocked with 2.25 cm of polyethylene plus a 3 mm thick boron panel or 11.5 cm of polyethylene, respectively. In addition, Lalonde (1997) reported that lining the maze walls with a neutron moderating material such as polyethylene was less effective in accelerator facilities.

2.4.5 *Direct-Shielded Door*

In some cases, the choice is made to save the space needed for a maze and use a heavy direct-shielding door (McGinley, 2001a). This door must have the same shielding value as the adjacent secondary barrier. The usual choice of shielding materials is a laminate of lead and steel (door shell) with the addition of BPE if photoneutrons are present. The boron concentration is most commonly 5 % by weight. These doors are very heavy. The practical limitation for a 120 cm (4 feet) wide door is in the range 8,000 to 9,000 kg for a swinging door. Beyond that weight, it is necessary to use two narrower doors

TABLE 2.1—*Dose-equivalent rate due to neutron capture gamma rays and neutrons at the outside maze entrance per unit absorbed-dose rate of x rays at the isocenter of a nominal 18 MV accelerator (McGinley and Miner, 1995).*[a]

Dose-equivalent rate (Sv h^{-1}) per unit absorbed-dose rate (Gy h^{-1}) of x rays at isocenter (Sv Gy^{-1})

Type of Maze and Door	Neutron Capture Gamma Rays	Neutrons	Total (neutrons plus gamma rays)
Conventional	5.8×10^{-7}	17.4×10^{-7}	23.3×10^{-7}
Reduced inner opening	2.6×10^{-7}	5.8×10^{-7}	8.4×10^{-7}
Inner boron door	1.9×10^{-7}	4.8×10^{-7}	6.7×10^{-7}
Inner BPE door	1.0×10^{-7}	1.5×10^{-7}	2.6×10^{-7}

[a]Neutron dose-equivalent rates were measured using a rem-meter calibrated with a heavy water-moderated ^{252}Cf neutron source.

or a sliding door. Sliding doors may be either hung from a rail or rolled on a steel support in the floor. This choice is an engineering decision and different manufacturers will have different opinions. In all cases these doors must be either electrically or hydraulically driven. It is imperative that plans be made for getting to the patient in the event of an electrical failure or failure of the driving mechanism. Given the weight of the door, it is likely that the most reliable approach is to make provisions for an escape hatch in one of the other barriers. Due to accidents with these door closure systems and failures of the support structures themselves, it is strongly recommended that periodic inspection and routine preventive maintenance programs be instituted for all doors that require an automatic opening and closing system (ACR, 2000). A typical room of this type is shown in Figure 2.9 with enlarged details of the door and frame in Figures 2.10 and 2.11.

2.4.5.1 *Design Problems with Direct-Shielded Doors.* The basic problem with these doors is illustrated in Figure 2.10. The three rays marked A, B and C are shown in the direction of leakage radiation from the isocenter. Ray A passes through reduced lead and BPE. Ray B passes through reduced concrete ~41 cm (16 inches) compared with the wall thickness of 104 cm (41 inches). Ray C is at

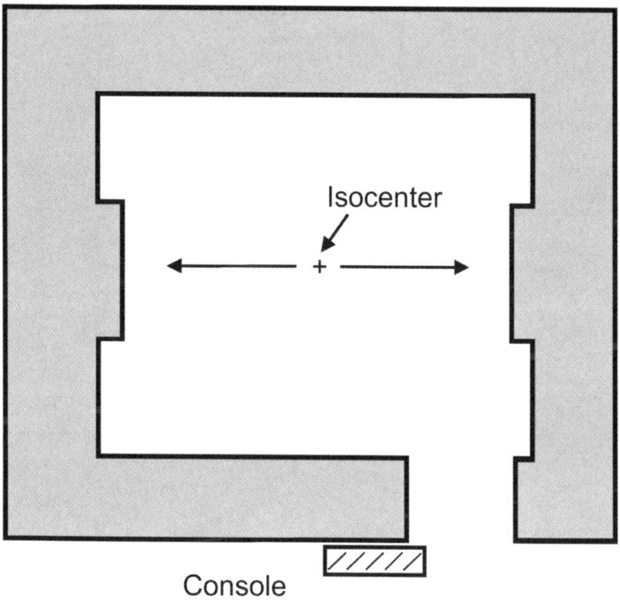

Fig. 2.9. Floor plan of room with direct shielding door.

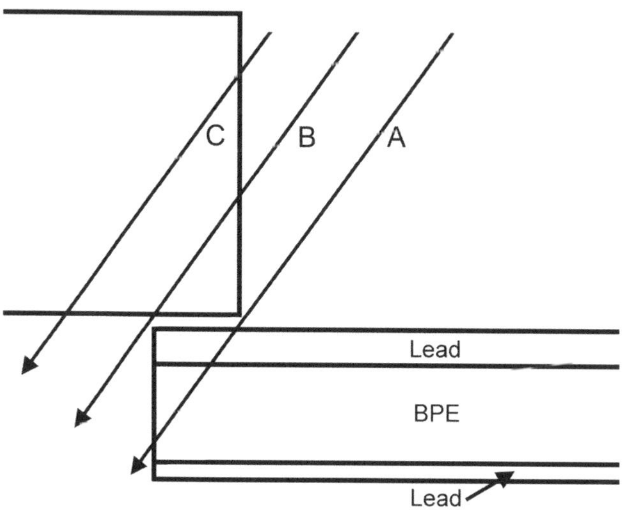

Fig. 2.10. Incomplete shielding at door.

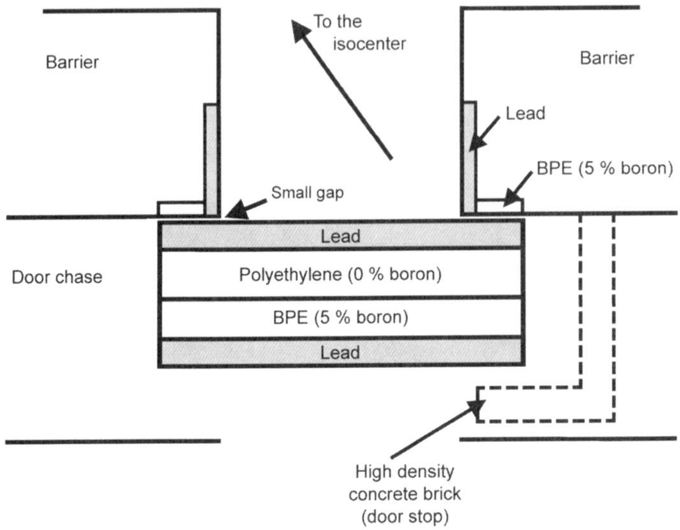

Fig. 2.11. Alternative to large overlap at door.

about the point where the slant thickness achieves the full thickness of the concrete of 104 cm. This is a general problem of direct-shielded doors and cannot be solved by juggling the position of the lead and BPE inside the door. There are two ways to solve this problem: one is to make the door overlap with the wall much larger; the other is to make a shielded doorstop as shown in Figure 2.11. In addition, lead and BPE may need to be added on the surface of the concrete wall. Making the door wider adds significantly to the weight and mechanical drive concerns as well as to the opening time and expense so it is a less desirable alternative. From the shielding viewpoint, these solutions are equally good and the choice may be made on architectural or other grounds. Since it is only possible to shield the gap (shielded door stop) between the door and the wall on the jamb side of the door, it is important to arrange the door so that the direct leakage radiation from the accelerator strikes the stop side rather than the side with the operator attached. Many of the practical problems encountered with these approaches are reviewed by McGinley (2001a) and must be considered in detail before a final design can be completed. The complexity of these doors is so great and the ramifications of mistakes are so significant that they should only be designed by persons with significant experience.

The preferred arrangement for the actual door shielding is with the lead on the accelerator room side and the BPE on the outside.

Some designers add lead on the outside to attenuate neutron capture gamma rays from the BPE. The neutron capture gamma rays from boron are only 480 keV and 1.9 cm (3/4 inch) of lead reduces the intensity by more than a factor of 100. The lead on the inside reduces the energy of the neutrons by inelastic scattering and this makes the BPE more effective. Again it is important that the leakage radiation from the accelerator strikes the door on the jamb side rather than the operator side since that is the easiest side on which to shield the gap.

2.4.5.2 *Neutron Capture Gamma Rays with Direct-Shielded Doors.* The question arises as to whether neutron capture gamma rays from the room surfaces need to be considered in choosing the lead thickness of a direct-shielded door. McGinley and Miner (1995) concluded that, since there are no known measurements of the neutron capture gamma-ray intensity inside a therapy room and the calculations are at best estimates, calculating the door shielding for leakage radiation and then adding 1 *HVL* is a conservatively safe approach. Note that the same provision does not have to be made for the secondary concrete walls since concrete neutron capture gamma-ray shielding will be conservatively safe if it is assumed that all neutron captures result in 7.2 MeV gamma rays. This implies a *TVL* in concrete of ~38 cm (15 inches), similar to that for leakage x rays (Table B.7). Thus adequate shielding for the one will produce a barrier that is adequate for the other.

2.4.5.3 *Alternative Room Design for Direct-Shielded Doors.* Barish (2005) describes a method of room design in which the therapy unit is positioned with the back of the gantry positioned on the door side of the room. A short stub wall is constructed to attenuate the leakage-radiation component of the secondary radiation that reaches the door. It is stated that this approach reduces the door thickness by approximately 50 %, and thus results in significant cost reductions as well as reductions in construction, operation and maintenance complexity.

3. Workload, Use Factor, and Absorbed-Dose Rate Considerations

The workload (W) is usually specified as the absorbed dose delivered to the isocenter in a week, and is selected for each accelerator based on its projected use. This includes the maximum number of patients (or fields) treated in a week and an estimate of the absorbed dose delivered per patient (or field). It *should* also include an estimate of the maximum weekly absorbed dose delivered during quality control checks, calibrations or other physics measurements. NCRP Report No. 49 (NCRP, 1976) recommended using $W = 1{,}000$ Gy week^{-1} for accelerators up to 10 MV, and NCRP Report No. 51 (NCRP 1977) recommended $W = 500$ Gy week^{-1} for higher energy accelerators. These workloads were recommended if a value for W could not be determined from direct knowledge of the accelerator use.

Treatments on clinical accelerators often involve using low- and high-energy x-ray beams and electron beams of various energies. For these dual-energy machines, the W at the higher energy will usually determine the shielding requirements. However, in some situations, to determine the required barrier thicknesses for both primary and secondary radiations it may be necessary to consider W separately for each x-ray beam quality. Workload for electron-beam operation can be disregarded, except for shielding accelerators with electron-beam-only operation, such as dedicated intraoperative units.

Technology and clinical applications have evolved since NCRP (1976) on shielding for radiation therapy vaults was prepared. For single-mode machines, the 6 MV accelerating voltage is commonly used, while dual x-ray mode machines often have a 6 MV low-energy beam and a 10 to 20 MV high-energy beam. Clinical electron beams may be an option on both single and dual x-ray mode machines. More recent clinical treatment procedures that may affect shielding requirements include SRS, TBI, and IMRT. In addition, there are new treatment devices that impact on these parameters such as tomotherapy (Mackie *et al.*, 1993) and robotic arms (Rodgers, 2005).

Total-body irradiation (TBI) procedures often require that the patient be located away from the isocenter by as much as 5 m, and therefore result in a large increase in the exposure to the barrier behind the patient as well as increased leakage-radiation production. When a multileaf collimator is used to modulate the photon fluence in order to improve treatment absorbed-dose distributions (*e.g.*, IMRT), the ratio of machine monitor units (MU) to absorbed dose delivered to the target volume may significantly increase above the nominal value of 1 MU cGy^{-1}. Leakage-radiation shielding requirements vary greatly with IMRT usage and the technology used to achieve the fluence modulation (Purdy, 2001). The appropriate workload and use-factor distributions to be used in the shielding calculations for such a facility require detailed deliberation beyond the conventional room design.

3.1 Conventional Procedures

When special procedures will not be used in the therapy room, a single workload (W) can be applied for the three principal sources or radiation: direct, leakage and patient-scattered. A single-use factor (U) distribution can also be used for these three sources of radiation.

3.1.1 *Conventional Workloads*

The weekly workload is determined by the number of patients treated per week and the average absorbed dose delivered per patient to the isocenter, and it is expressed in gray per week as measured at 1 m from the x-ray target. Note, however, that since the workload includes the gray per week ancillary to the actual treatment of patients (*e.g.*, QA, maintenance, and physics developmental activities), it is not just equal to the gray per patient times the number of patients per week even though it is stated as an average gray per patient per week.

NCRP Report No. 49 (NCRP, 1976) recommended using W = 1,000 Gy $week^{-1}$ for accelerators up to 10 MV, and NCRP Report No. 51 (NCRP, 1977) recommended W = 500 Gy $week^{-1}$ for higher energy accelerators. These workloads were recommended for use if a value for W could not be determined from direct knowledge of the accelerator use. A survey by Kleck and Elsalim (1994) found that accelerators operating at 6 MV had a workload <350 Gy $week^{-1}$ and that the workload at the high energy of dual-energy accelerators was <250 Gy $week^{-1}$. These values are comparable to those in a later report (Mechalakos *et al.*, 2004), which gave values of 450 and 400 Gy week^{-1} for a 6 MV single-energy and a 6 and 15 MV dual-energy machine, respectively.

Typically, the proportion of high-energy x-ray usage on a dual-energy machine ranges from 40 to 70 % (monitor unit weighted). Allowance for intended special use or dedicated patient type *shall* be considered in determining the mix of high- and low-energy x-ray usage.

The average number of patients treated per day *shall* be estimated conservatively high based on factors such as the number of treatment machines in the total facility, historical factors, and demographics. With the advent of computerized verify and record systems, it is now possible to collect data for the beam-on time that are truly representative of the workload and use factor for various gantry angles and energies. These data *should* also be monitored after the initiation of treatments in a newly shielded facility to compare to the design assumptions and decide on the need for future intervention or remediation. In lieu of such specific information, recommended patient loads based on survey averages for a busy facility are given as 30 patients per day (Mechalakos *et al.*, 2004).

3.1.2 *Conventional Use Factors*

The use factor as a function of the gantry angle [$U(G)$] gives the fraction of the weekly workload for which the gantry or beam is oriented in an angular interval centered about angle G. It is to be noted that the distribution of U values as a function of G depends on the size of angular interval represented. Table 3.1 presents use-factor distributions for standard treatment techniques (*i.e.*, omitting procedures that significantly alter the use-factor distribution such as SRS and TBI) for 90 and 45 degree binning of the same machine data.[13]

Clearly, when viewed in 90 degree bins these data agree with the expected traditional "four field" approach to treatment. However, upon closer inspection, some of the features that may be unique to a facility begin to appear. For example, a large fraction of tangential breast fields would significantly affect the use factor for gantry angles toward the wall-floor and wall-ceiling interfaces. It is very important that these be considered, especially if any tapering of the barrier thickness is used to account for the beam obliquity.

[13]Rodgers, J.E. (2001). Personal communication (Georgetown University, Washington). Unpublished reanalysis of the survey data in Kleck and Elsalim (1994).

TABLE 3.1—*High-energy (dual x-ray mode) use-factor distribution at 90 and 45 degree gantry angle intervals.*[a]

Angle Interval Center	U (%)
90 degree interval	
0 degree (down)	31.0
90 and 270 degrees	21.3 (each)
180 degrees (up)	26.3
45 degree interval	
0 degree (down)	25.6
45 and 315 degrees	5.8 (each)
90 and 270 degrees	15.9 (each)
135 and 225 degrees	4.0 (each)
180 degrees (up)	23

[a]Rodgers, J.E. (2001). Personal communication (Georgetown University, Washington). Unpublished reanalysis of the survey data in Kleck and Elsalim (1994).

3.2 Special Procedures

Due to the increasing sophistication of treatment techniques, it is often not possible to use single estimates for the values of the workload and use factor in the shielding design. Rather the nuances of the different proposed procedures must be considered separately and then in combination in order to assess the total impact on the shielding design.

3.2.1 *Total-Body Irradiation Considerations*

When total-body irradiation (TBI) is part of the clinical practice there are several important concerns that *shall* be addressed (Rodgers, 2001). The workload (Gy week^{-1} at 1 m) for TBI is usually significantly higher than conventional radiotherapy treatments for the same unit absorbed dose to the patient, because extended treatment distances are used. The workload for leakage radiation is also higher for the same reason, and is different from the workload for patient-scattered radiation and wall-scattered radiation. The TBI workload (W_{TBI}) is the absorbed dose at 1 m, and therefore is the product of the weekly total TBI absorbed dose to the patient (D_{TBI}) and the square of the TBI treatment distance (d_{TBI} in meters), as shown in Equation 3.1.

$$W_{TBI} = D_{TBI}\, d_{TBI}^2 \tag{3.1}$$

Since the patient is usually positioned close to one of the primary barriers for the TBI procedure, instead of at the isocenter, the effect of scattered radiation at the room entrance needs to be considered. This is especially true if there is a maze instead of a direct-shielded door. In some room arrangements, the source of scattered radiation (the TBI patient and wall behind patient) will be much closer to the room entrance than to the isocenter, and the consequent dose-equivalent rate at the entrance will be higher.

Distinct workloads need to be determined for the TBI primary-barrier radiation and the leakage-radiation contributions to secondary barriers that will be different from the workload applicable to patient-scattered radiation to secondary barriers. Consider, for example, a weekly workload without TBI that is 300 Gy week^{-1} at 1 m. If an average of one TBI patient is to be treated per week with 12 Gy at a location 4 m from the x-ray target, the TBI irradiation increases the workload at 1 m by 192 Gy week^{-1}. This TBI procedure workload contribution affects the various radiation processes (primary, scatter and leakage) differently. The total dose equivalent at a shielded location beyond a primary barrier is determined from the sum of dose contributions at that location from all accelerator usage. Thus, the TBI workload increases the dose equivalent to the primary barrier behind the patient. It also increases the leakage-radiation contribution to all barriers. Scattered radiation from the isocenter to secondary barriers is not changed.

For the above example where only conventional and TBI procedures contribute to the location behind the TBI barrier, the conventional workload use factor is estimated (Table 3.1) to be 0.21 whereas for TBI procedures restricted to this one barrier the TBI use factor is one. Thus, the primary-radiation barrier workload at isocenter (1 m) directed toward the TBI barrier is 255 Gy week^{-1} [300 Gy week^{-1} (0.21) + 192 Gy week^{-1}]. This is also the workload applicable to scattered radiation from the TBI barrier and from the TBI patients positioned near that barrier. Scattered radiation from non-TBI patients at the isocenter to all secondary barriers is determined by the conventional workload, 300 Gy week^{-1}. All secondary barriers, in this example, have leakage-radiation contributions determined from a leakage-radiation workload at 1 m of 492 Gy week^{-1} (300 Gy week^{-1} + 192 Gy week^{-1}). More general situations are considered in Section 3.2.5.

3.2.2 *Intensity Modulated Radiation Therapy Considerations*

Intensity modulated radiation therapy (IMRT) procedures (as well as SRS and SRT) often use small beams produced by multileaf collimators or cones with narrow openings. Due to the small field sizes of the large number of beamlets used, the total accelerator monitor units (MU) required are much higher than would have been required for conventional radiotherapy for the same absorbed dose to the patient. The ratio of the average monitor unit per unit prescribed absorbed dose needed for IMRT (MU_{IMRT}) and the monitor unit per unit absorbed dose for conventional treatment (MU_{conv}) is called the IMRT factor.

The IMRT factor (C_{I}) may be obtained using the following method. Take a sample of IMRT cases and calculate the average total monitor unit required to deliver a unit prescribed absorbed dose per fraction (D_{pre}) for the "i" cases (include the required QA measurements for each case averaged over the total number of treatment fractions). Then MU_{IMRT} is given by:

$$MU_{\text{IMRT}} = \Sigma_{\text{i}} \frac{MU_{\text{i}}}{(D_{\text{pre}})_{\text{i}}} \tag{3.2}$$

Then calculate the monitor unit required to deliver the same unit absorbed dose to a phantom at 10 cm depth at 100 cm source-to-isocenter distance, using field size 10 × 10 cm, to obtain the quantity MU_{conv} (a depth of 10 cm is chosen to represent an average treatment depth for conventional radiotherapy). The IMRT factor C_{I} is simply equal to MU_{IMRT} divided by MU_{conv}:

$$C_{\text{I}} = \frac{MU_{\text{IMRT}}}{MU_{\text{conv}}} \tag{3.3}$$

Values of C_{I} can range from 2 to 10 or more (Followill *et al.*, 1997; Mutic and Low, 1998; Purdy *et al.*, 2001; Rodgers, 2001). The increase in monitor units required by IMRT does not significantly increase the workload for the primary barrier, or for the patient- and wall-scattered radiation components of the secondary barrier. This is because the absorbed doses to the patient for IMRT and conventional radiotherapy are similar. Hence, the direct workload at

1 m from IMRT procedures (W_{IMRT}) and from conventional procedures (W_{conv}) are equal regardless of whether the patients were treated with or without IMRT. However, the contribution to the leakage-radiation workload (W_L) is significantly higher by a factor as large as the IMRT factor C_I, depending on the fraction of patients treated with IMRT (see Section 3.2.5.3 and the examples in Section 7, as well as Equations 2.8, 2.10, and 2.12).

3.2.3 *Quality Assurance*

If quality assurance (QA) measurements are routinely performed during normal working hours, the effect on the workload may not be negligible compared with the conventional treatment workload, and the impact on barrier shielding requirements should be evaluated. Quality assurance measurements include daily, monthly and annual tests, commissioning measurements, IMRT absorbed-dose verifications, research, and other activities conducted with the radiation beam on. If IMRT absorbed-dose verification constitutes a significant part of the QA measurements, a QA factor (C_{QA}) similar to C_I may be required to account for the increase in monitor units.

3.2.4 *Dedicated Purpose Machines*

Care must be taken when designing special purpose vaults where predominantly one type of patient treatment will be given. For example, treatments of mostly SRS result in more uniform $U(G)$ distributions when compared to traditional methods, while breast treatments that employ physical wedges and some IMRT treatments may significantly affect not only the monitor unit efficiency ($C_I > 1$) but also may have very nonuniform $U(G)$ distributions. In addition, there are some special purpose machines that present unique situations, as discussed in Section 5. Given the evolving nature of radiation technology and methods, and the uniqueness of some dedicated facilities, a cost-benefit analysis of future alterations in the planned usage *should* be considered.

3.2.5 *Effect of Special Procedures on Shielding Calculations*

This Section treats more specifically how various special procedures affect the shielding calculation details, especially as they affect the workload and use factor values to be used in the equations of Section 2.

3.2.5.1 *Primary-Barrier Calculations.* Equations 2.1 and 2.9 in Section 2 may be used to determine the effect of special procedures on those barriers that are determined (at least partially) by the primary beam. To account for the differences in usage and workload as described above, the products of workload and use factor in Equations 2.1 and 2.9 are replaced by the sum of the products of the workload times use factor for each technique (*i.e.*, conventional treatments, TBI, IMRT, QA, and other nonconventional treatment procedures and activities) as in Equation 3.4 (Rodgers, 2001).

$$WU]_{pri} = WU]_{wall\ scat} \tag{3.4}$$

$$= (W_{conv}\ U_{conv} + W_{TBI}\ U_{TBI} + W_{IMRT}\ U_{IMRT} + W_{QA}\ U_{QA} + ...)$$

In Equation 3.4:

$WU]_{pri}$ and $WU]_{wall\ scat}$ = workload-use factor products for the primary radiation barrier and the wall-scattered-radiation barrier, respectively
W_x = workload in gray per week at 1 m for procedure type x (*e.g.*, conventional, TBI, IMRT, QA)
U_x = use factor or fraction of time that the beam is likely to be incident on the barrier for procedure type x

3.2.5.2 *Patient- or Phantom-Scattered-Radiation Calculations.* The shielding required for the patient- or phantom-scattered radiation component for the secondary barriers is evaluated with Equation 2.7. The workload will be given by the sum of the workloads for all techniques performed at the isocenter (Equation 3.5).

$$W]_{pat\ scat_{iso}} = W_{conv} + W_{IMRT} + W_{QA} + \cdots \tag{3.5}$$

Typically, a scattering angle of 90 degrees and a use factor of unity are assumed which gives a conservatively safe result, since the increased intensity of small angle scatter relative to 90 degree scatter is generally offset by the much smaller use factor for the gantry angles producing the small angle scatter.

For TBI performed at extended source-to-surface distance the patient-scattered radiation arises from a different location than the isocenter. Consequently, the values d_{sca} and d_{sec} will be different. The shielding required for the TBI component of patient-scattered radiation, therefore, needs to be determined separately with Equation 2.7. Note that for maze and door calculations the scattered radiation from the barrier beyond the patient is also significant as discussed in Section 3.2.5.4.

If the *TVL*s required to shield the two components of patient-scattered radiation (patient at the isocentric or at the TBI position) differ by <1 *TVL*, then the larger *TVL* is used and a further *HVL* of shielding is added. Otherwise the larger *TVL* value is used.

3.2.5.3 *Leakage-Radiation Shielding Considerations.* For leakage-radiation considerations, Equation 2.8 may be used. The workload W_L is given by Equation 3.6.

$$W_L = W_{conv} + W_{TBI} + C_I\ W_{IMRT} + C_{QA}\ W_{QA} + \ldots \qquad (3.6)$$

In Equation 3.6, W_{TBI} is determined from Equation 3.1 and C_I is determined as discussed in Section 3.2.2. W_L is used for photon leakage and photoneutron production calculations.

3.2.5.4 *Maze Entrance Calculations.* The method for evaluating the maze entrance dose equivalent is described in Section 2.4. In this method, Equations 2.9 through 2.12 are used in order to obtain the dose equivalents at the maze entrance due to: wall-scattered radiation after attenuation by the patient, head-leakage scattered radiation, patient-scattered radiation, and leakage-radiation transmission through the inner maze wall, respectively.

For special procedures, the dose equivalent from radiation scattered by the patient is determined separately for isocentric and nonisocentric (TBI) techniques. In Equation 2.11, *W* will take the value of $W]_{pat\ scat_{iso}}$ from Equation 3.5 for isocentric techniques; while, for TBI, the *W* value will be W_{TBI} from Equation 3.1 and the distances will need to be measured based upon the geometry of the patient's position within the room for the TBI treatments. The dose equivalent arising from patient-scattered radiation at the maze entrance will be given by Equation 3.7.

$$H_{ps} = H_{ps,iso} + H_{ps,TBI} \qquad (3.7)$$

In Equation 3.7, $H_{ps,iso}$ and $H_{ps,TBI}$ are the dose equivalent from radiation scattered by the patient for isocentric and TBI techniques, respectively. The energies of these contributions may be significantly different. In Equations 2.9 and 2.11, $W\ U_G$ will take the value of $W\ U_{pri}$ in Equation 3.4.

The dose equivalents at the maze entrance due to head-leakage radiation scattered by the end wall of the maze, and the head-leakage radiation transmitted through the maze wall may be obtained from Equations 2.10 and 2.12, respectively, using the workload W_L given by Equation 3.6. The assumption represented in Equation 2.14 by the coefficient 2.64 might be invalid for these special procedures (such as TBI) since the use factor is highly dependent on the particular special procedure. In that case, the maze entrance dose equivalents are evaluated through repeated application of Equations 2.9 through 2.12 for each of the principle primary barriers, and the results for the barriers are summed to obtain the total dose equivalent at the maze entrance.

For higher energy machines (>10 MV), the total dose equivalent at the maze entrance is dominated by the neutron capture gamma-ray dose equivalent as determined from Equations 2.15, 2.16, and 2.17, and the neutron dose equivalent as determined from Equations 2.19, 2.20, and 2.21. For special procedures like TBI, IMRT, and QA, the workload W in Equations 2.17 and 2.21 is the leakage-radiation workload W_L from Equation 3.6. However, the dose equivalents obtained using Equations 2.17 and 2.21 are for radiation beams pointing downward. Thus, for example, if a TBI setup employs a horizontal beam orientation with the gantry head closest to the inner maze entrance, the actual neutron dose equivalent will be higher by roughly the square of the ratio of the distances from the inner maze entrance to the head of the accelerator for the downward and horizontal positions. Therefore, in order to be conservatively safe in the design of rooms that will employ special procedures, it is recommended that, for high-energy beams, the dose equivalents (obtained from Equations 2.17 through 2.21) *should* be multiplied by 1.5.

At the present time, many IMRT procedures are performed using energies <10 MV. At these energies, the neutron capture gamma-ray and neutron dose equivalents in the room and the maze entrance are usually low, and the increase in dose equivalent at the maze entrance is entirely due to the increase in leakage radiation because of the C_I-fold rise in monitor units. However, if energies of 10 MV and higher are used, the neutron capture gamma-ray and neutron dose equivalents at the maze entrance will increase rapidly with energy and may dominate over the leakage-radiation components.

3.3 Time Averaged Dose-Equivalent Rates

When designing radiation shielding barriers it is usual to assume that the workload will be evenly distributed throughout the year. Therefore, it is reasonable to design a barrier to meet a weekly value equal to one-fiftieth of the annual shielding design goal (NCRP, 2004). However, further scaling the shielding design goal to shorter intervals is not appropriate and may be incompatible with the ALARA principle. Specifically, the use of a measured instantaneous dose-equivalent rate (*IDR*), with the accelerator operating at maximum output, does not properly represent the true operating conditions and radiation environment of the facility. It is more useful if the workload and use factor are considered together with the *IDR* when evaluating the adequacy of a barrier. For this purpose, the concept of time averaged dose-equivalent rate (TADR) is used in this Report along with the measured or calculated *IDR*.

The TADR is the barrier attenuated dose-equivalent rate averaged over a specified time or period of operation. TADR is proportional to *IDR*, and depends on values of *W* and *U*. There are two periods of operation of particular interest to radiation protection, the week and the hour.

3.3.1 *Weekly Time Averaged Dose-Equivalent Rate*

The weekly time averaged dose-equivalent rate (R_w) is the TADR at the specified location averaged over a 40 h workweek. For primary barriers it is given by Equation 3.8.

$$R_W = \frac{IDR\ W_{pri}\ U_{pri}}{\dot{D}_o} \tag{3.8}$$

In Equation 3.8:

R_w = TADR averaged over one week (Sv week^{-1})

IDR = instantaneous dose-equivalent rate (Sv h^{-1}) measured with the machine operating at the absorbed-dose output rate $\dot{D}_o$. *IDR* is specified at 30 cm beyond the penetrated barrier, and for accelerator measurements it is averaged over 20 to 60 s depending on the instrument response time and the pulse cycle of the accelerator

$\dot{D}_o$ = absorbed-dose output rate at 1 m (Gy h^{-1})

W_{pri} = primary-barrier weekly workload (Gy week^{-1})

U_{pri} = use factor for the location

For secondary barriers, R_w has contributions from both leakage and patient-scattered radiation, shown as in Equations 3.9 and 3.10.

$$R_W = \left(IDR_L \frac{W_L}{\dot{D}_o} \right) + \left(IDR_{ps} \frac{W_{ps}\ U_{ps}}{\dot{D}_o} \right) \tag{3.9}$$

$$IDR_{ps} = IDR_{total} - IDR_L \tag{3.10}$$

In Equations 3.9 and 3.10:

IDR_L = dose-equivalent rate measured at point located 30 cm beyond the secondary barrier and in the *absence* of a phantom at the isocenter

IDR_{total} = dose-equivalent rate measured at the same point in the presence of a phantom

IDR_{ps} = dose-equivalent rate at the same point due to patient-scattered radiation

The concept of R_w is useful in the evaluation of barrier adequacy as illustrated in examples in Section 7.

3.3.2 *In-Any-One-Hour Time Averaged Dose-Equivalent Rate*

Some radiation safety regulations specify a limit for the time averaged dose-equivalent rate (TADR) based on the integrated dose equivalent in-any-one-hour (R_h) in uncontrolled areas. For example, the U.S. Nuclear Regulatory Commission (NRC) specifies that the dose equivalent in any unrestricted area from external sources not exceed 0.02 mSv in-any-one-hour (NRC, 2005a). R_h derives from the maximum number of patient treatments that could possibly be performed in-any-one-hour when the time for setup of the procedure is taken into account. Thus:

$$R_h = N_{max}\ \overline{H}_{pt}\ . \tag{3.11}$$

In Equation 3.11:

N_{max} = maximum number of patient treatments in-any-one-hour with due consideration to procedure set-up time

$\overline{H}_{pt}$ = average dose equivalent per patient treatment at 30 cm beyond the penetrated barrier

But $\overline{H}_{pt}$ is also equal to the time averaged dose equivalent per week (R_w) divided by the average number of patient treatments per week ($\overline{N}_w$):

$$\bar{H}_{pt} = \frac{R_w}{\bar{N}_w} \tag{3.12}$$

and

$$\bar{N}_w = 40\ \bar{N}_h \tag{3.13}$$

In Equations 3.12 and 3.13:

$\bar{N}_h$ = average number of patient treatments per hour
40 = 40 work hours per week.

Therefore:

$$R_h = \frac{N_{max}\ R_w}{\bar{N}_w} = \left(\frac{N_{max}}{40\ \bar{N}_h}\right) R_w = \left(\frac{M}{40}\right) R_w \tag{3.14}$$

In Equation 3.14:

$M = N_{max} / \bar{N}_h$, and M is always ≥ 1.

Here R_h, in units of sievert, is the TADR to be compared to the dose-equivalent limit mentioned above of 0.02 mSv in-any-one-hour outside the barrier.[14]

The average number of patient treatments per hour ($\bar{N}_h$) is calculated by dividing the weekly workload by the product of the average absorbed dose at the isocenter per patient treatment and 40 work hours per week. Alternately, Equation 3.13 can be used, in which $\bar{N}_h = \bar{N}_w / 40$.

In some circumstances, for example a dedicated stereotactic radiosurgery machine used only a few times per day, N_{max} may be much larger than $\bar{N}_h$, and the values of R_h determined for evaluation of shielding of public areas may not be satisfactory. In this situation, it may be necessary to restrict N_{max}, and/or the minimum occupancy factor assigned to an area, and it may be necessary to add additional shielding. In any event, N_{max} *shall* be based on a realistic estimate that includes the set-up time.

Some example calculations of TADRs are given in Section 7.

[14] R_h is not the shielding design goal (P). It is a separate requirement, in some regulations, for the upper bound of the dose-equivalent rate in-any-one-hour. The shielding design goal is discussed in Section 1.4.

4. Structural Details

4.1 General

The shielding of the radiation treatment room *shall* be constructed so that the protection is not compromised by joints, openings for ducts, pipes, etc. passing through the barriers, or by conduits, service boxes, etc. embedded in the barriers. Door design for high-energy machines also requires special consideration to ensure adequate protection without sacrifice of operational efficiency. These and related problems are considered in detail in this Section.

It is important that the shielding is designed and installed properly; corrections made after the room is completed can be expensive. It is often impractical to make an overall experimental determination of the adequacy of the shielding prior to the completion of the building construction and the installation of the radiation equipment. Thus, periodic ongoing inspection during the entire construction period *should* be performed. Sometimes properly constructed shielding is compromised by subsequent changes that are made to install ducts, recessed boxes, and other hardware or changes made to walls, roof or floors.

There is considerable variation in the shielding requirements for therapy installations due to the wide range of energies and different types of equipment and clinical techniques used. Careful attention to detail before, during and after construction, may result in appreciable savings, particularly in the energy range where the shielding is very costly. Additional details concerning equipment design and use can be found in NCRP Report No. 102 (NCRP, 1989).

As a note of caution, the construction industry in the United States still uses "traditional" units almost exclusively. Thus, if the shielding designer does the calculation in metric units, the designer is well advised to also make the conversion to the traditional units rather than leave it to the architect or contractor.

4.1.1 *Location*

Operational efficiency, initial cost, as well as provision for future expansion or increased workload, *should* be considered when locating a therapy installation. Proximity to adjunct facilities, ready access for inpatients and outpatients, and consolidation of all therapeutic radiological services, however, may be more important than construction cost.

Radiotherapy facilities are often constructed below ground level as the most economical strategy to minimize shielding costs associated with exterior walls. For rooms below ground, the reduction in shielding costs for floors and outside walls needs to be weighed against the expenses of excavation, watertight sealing, and providing access. For rooms on or above ground, the outside walls almost always require shielding, using concrete, brick, stone or other suitable material. The range of potential attenuating properties of these materials is very wide and the qualified expert *should* request specific exterior wall design specifications from the architect prior to determining the shielding design.

4.1.2 *Provision for Future Needs*

Clinical procedures such as TBI or IMRT could significantly alter the workload, use factor, and even the physical size of the room. Cost and inconvenience of future alterations may be reduced by providing extra rooms initially or by allowing for future enlargement of rooms to accommodate replacement equipment of larger size, higher energy, and with increased workload. If the installation is on an upper floor, room enlargement or contiguous expansion may be impossible. If the installation is on the ground floor, expansion onto the surrounding grounds may be most economical, requiring shielding only for walls, and possibly the ceiling, without floor shielding. Expansion over an occupied area may require extra structural support and floor shielding. Expansion underground may require additional excavation, possibly with relocation of sewerage and other services. Future need for additional services (*e.g.*, electrical, water, vacuum, oxygen) *should* be considered during initial planning.

4.1.3 *Size of Treatment Room*

The desirable size of a treatment room depends upon the type of the therapy equipment, the type of treatments, and the use of special equipment for research and teaching. Procedures like TBI and IORT require larger size rooms. Making the room larger than necessary may permit the installation of additional ancillary equipment, or replacement of the original therapy equipment by larger equipment. Specific vendor requirements will affect the placement of primary barriers, ceiling heights, and maze widths. However, larger rooms are more costly to build, require more walking for the staff, and encourage use for extraneous purposes. A room larger than necessary will also increase the amount of time and effort to clean and prepare when used for IORT procedures (Palta *et al.*, 1995).

The entrance and inner maze openings *should* be adequate for transport of equipment and patients in and out of the room, but small enough to minimize scattered radiation to the maze door.

4.1.4 *Interlocks and Warning Lights*

Interlocks and warning lights *shall* be provided so that the radiation beam status is observable from both inside the treatment room and at the control console (NCRP, 1989). The output of the therapy machine may be so high that a person who is accidentally in the treatment room when the machine is turned "on" may receive an excessive exposure during the time required to reach an access door. This hazard can be reduced by having "emergency-off" buttons at appropriate positions inside the treatment room, that, when pressed, will terminate the irradiation.

4.1.5 *Control Console*

The control console *shall* be located outside the treatment room and it *should* be beyond a secondary barrier in order to keep the console shielding thicknesses to a minimum and to achieve radiation doses to staff that are consistent with the ALARA principle. The entrance to the treatment room *should* be visible from the console and there *shall* be provision for both visual and audio communication with the patient.

4.2 Barriers

Primary protective barriers for megavoltage installations will usually be composed of concrete several meters thick, or the equivalent thickness of other materials that provide equivalent shielding. Therefore, to ensure economical construction, careful consideration of room location, radiation beam orientation, and use is necessary.

Restricting the beam orientation will limit the area requiring primary protective barriers to that which can be struck by the useful beam plus a suitable margin (Section 2.2.2). In the design of primary protective barriers, due consideration *should* be given to possible future changes in techniques that may increase the use factor and/or workload. The location of primary barriers will also be affected by the types of occupied spaces beyond the barrier. For equipment capable of rotating the radiation beam towards barriers not shielded for primary radiation, mechanical or electrical means *shall* be provided to prevent the useful beam from striking those barriers.

When the vault is below ground level, the projection of the beam through the wall at the level of the ground surface *shall* be given special consideration (Figure 4.1) to ensure that the upper edge of the largest projected beam will be far enough below the surface to prevent significant scattered radiation from reaching the surface (Section 2.2.1).

Protective source housings for megavoltage equipment frequently provide more shielding than the recommended minimum specified in NCRP Report No. 102 (NCRP, 1989). Unless it is established that such extra shielding is present, the structural shielding design *shall* be based on the assumption that only the recommended minimum is provided. If pertinent data on leakage radiation are available, however, appreciable reduction in secondary-barrier requirements may be possible.

When the electron beam of an accelerator is not properly directed to or focused on the target (such as in the use of a research beam), it may strike other parts of the equipment and result in a substantial increase in the leakage radiation. This possibility *should* be considered in the shielding design and in the requirements for monitoring.

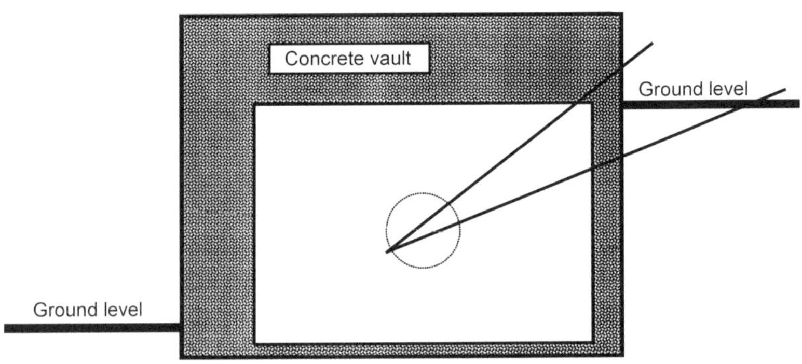

Fig. 4.1. Elevation of vault with partial shielding at ground level.

4.3 Shielding Materials

The shielding for electron accelerators of higher energy should take into account both photons and neutrons. Neutron shielding requires materials that contain hydrogen, while x-ray shielding materials need mass and high atomic number. Separate materials can be used for the two purposes, or materials that are good shields for both neutrons and x rays can be used. The materials that have been used for this purpose in order of frequency are: ordinary concrete, heavy concrete, lead, steel, polyethylene or paraffin, earth, and wood. Each will be discussed separately in Section 4.3 and Table B.3 summarizes some of the relevant radiation shielding properties for several of these materials.

4.3.1 *Ordinary Concrete*

Concrete has many advantages and is the most commonly used material. It can be poured in almost any configuration and provides good x-ray shielding and structural strength as well as neutron shielding. It is relatively inexpensive and the formwork is well over half the total cost so that it is often cheaper to use more concrete than to construct elaborate forms. Ordinary concrete is usually considered to have a density of 2.35 g cm^{-3} (147 lb $feet^{-3}$). This density for concrete is not always easy to obtain. It is usually cost effective to arrange for the service of an independent testing company to verify the density of each truckload of concrete. In some parts of the country such as Florida, local sand and aggregate results in concrete of much lower density, as low as 2 g cm^{-3} (125 lb $feet^{-3}$) (Rogers *et al.*, 1995), and that can have a significant affect on the shielding properties of the concrete. Walker and Grotenhuis (1961) list nine different concretes with densities ranging from 2.09 to 2.50 g cm^{-3} that are considered ordinary concretes. Hydrogen content ranged from 0.8 to 2.4×10^{22} atoms cm^{-3}. However, hydrogen content does not vary that much in practice and it is not necessary to consider the hydrogen variability for neutron shielding of electron accelerators, since the shielding thickness for an all-concrete medical-use accelerator room is determined by the x-ray shielding requirements rather than the neutron shielding requirements (IAEA, 1979).

4.3.2 *Heavy Concrete*

Heavy concrete is often used in construction of radiation therapy rooms. Any concrete with a density of >2.35 g cm^{-3} (147 lb $feet^{-3}$) can be considered heavy concrete. Heavy concretes are usually used where space is at a premium. The increased density is achieved by adding various higher density aggregates to concrete to increase photon attenuation. Common among these are iron ores and minerals such as haematite or ilmenite, barium minerals such as barytes, or ferrous materials. The density can range up to 5.2 g cm^{-3} for iron ores, 4.4 g cm^{-3} for barium minerals, and over 7 g cm^{-3} for ferrous materials.

The disadvantages of heavy concretes are cost and handling difficulties. Few contractors are accustomed to working with heavy concretes and are properly concerned about problems such as settling out of the heavy aggregates, structural strength and handling difficulties. Concrete pumps are not designed for such dense material and a concrete truck capable of carrying nine cubic yards of ordinary concrete may be able to carry only half that much heavy concrete safely. There are pre-cast interlocking heavy concrete blocks available commercially which mitigate many of these problems. Nominal densities of 3.8 and 4.8 g cm^{-3} (240 and 300 lb $feet^{-3}$) are available. The higher densities are achieved using higher-Z aggregate or by embedding small pieces of scrap steel or iron in cement. A commercial version of such a material has been made in the form of interlocking blocks with a density of 4.8 g cm^{-3} (Barish, 1993).

The advantage of using these heavy concretes is that primary walls can be made thinner. If a primary barrier is made with this material and the secondary barrier with ordinary concrete, a uniform thickness can be achieved for the wall. However, although the *TVLs* for these materials are less for photons, they are not necessarily less for neutrons since the additives are high-Z compounds with relatively low neutron absorption capability. In this case, it can no longer be assumed that a primary shielding barrier adequate for photons will be adequate for neutrons, and radiation protection surveys *shall* be performed to confirm that the primary shielding barrier is adequate also for neutrons.

On a more practical side are the questions of availability and cost. Heavy concretes made locally at the construction site are not subject to any industrial standards or consistency specifications and therefore need to be checked by the individual customer. Those that are prefabricated can be made to more rigorous standards and tested in advance, however they are more costly to ship to the construction site and are supplied by only a few companies. In either case, the cost of heavy concrete is higher than for ordinary concrete. On the other hand, the total "footprint" of the treatment room is greatly reduced and this may be considered a more important factor in some sites as well as cost savings in the long run when the reduced square footage requirement is amortized.

4.3.3 *Lead*

Lead has a very high density of 11.35 g cm^{-3} (708 lb $feet^{-3}$) and is an excellent shielding material for x and gamma rays where space is at a premium. Lead is available in a variety of forms such as bricks, sheets and plates. Since lead is malleable, it will not support its own weight and therefore requires a secondary support structure. It is nearly transparent to fast neutrons but does decrease the energy of higher-energy neutrons by inelastic scattering. However, below ~5 MeV, the inelastic scattering cross section for neutrons drops sharply. Because of its toxicity, lead *should* be encased in concrete or be protected by heavy coats of paint or drywall.

4.3.4 *Steel*

Steel is also relatively expensive (in comparison to concrete) but is not toxic (in comparison to lead). Its density is ~7.8 g cm^{-3} (~490 lb $feet^{-3}$) though it is commonly found in various alloys rather than as pure iron. Its shielding value for x and gamma rays is intermediate between lead and concrete. It is also nearly transparent to neutrons but does reduce the neutron energy by inelastic scattering with the inelastic scattering cross section dropping sharply below 1 MeV. Steel is a good structural material. When building laminated metal-concrete shields in a new facility, steel is usually less expensive than lead. When retrofitting an existing room, lead is usually less expensive.

4.3.5 *Polyethylene and Paraffin*

These two materials will be considered together since they are very similar. Paraffin, sometimes called paraffin wax, has the same percentage of hydrogen (14.3 %) as polyethylene and is less expensive. However, it has lower density and is flammable so it is usually avoided in any permanent barriers. Polyethylene is perhaps the best neutron shielding material available, but it is relatively expensive. It is available both pure and loaded with varying percentages of boron to increase the thermal neutron capture. It is usually used for neutron shielding, and where space is at a premium (*e.g.*, in doors) around ducts or in ceiling shields where insufficient height is available for less costly methods. It is easy to fabricate and relatively strong. Standard BPE contains 5 % boron by weight.

4.3.6 *Earth*

Earth is also commonly used as a shielding material by placing the accelerator room partially or entirely underground. Earth is not a well-defined material and its density can vary considerably. While earth composition is quite variable, it is not too different elementally from concrete and it is sufficient to consider it as equivalent to concrete with a density of 1.5 g cm^{-3} (95 lb $feet^{-3}$). A common problem with a room that is below grade is that there may be a diagonal path up through mostly earth where the shielding is insufficient (Figure 4.1). Sand that is in contact with the earth will normally be damp enough that it can be treated the same as earth. Dry sand, however, such as bagged sand, needs some hydrogenous material beyond it in order to be an adequate neutron shield.

4.3.7 *Wood*

Wood is inexpensive, readily available, and easy to fabricate but has low density. Wood varies from one kind to another, but for shielding purposes can be considered cellulose with a hydrogen density of ~6 %. It can be useful for temporary shielding (*e.g.*, as a temporary door with roughly 4 cm of wood equivalent to 1 cm of polyethylene). Fire resistant plywood contains boron and is a good thermal neutron shield.

4.3.8 *Rebar and Form Ties*

The subject of rebar and form ties is seldom mentioned in the shielding literature. This is in part because almost all of the measurements for both ordinary and heavy concrete are based upon small samples, often <15 cm diameter, and thus the concrete contained no rebar. It is also partly because rebar placement and size are based on structural requirements and vary quite a bit. Almost all concrete shielding data are based on concrete without rebar and form ties, so the use of rebar and its effects on shielding has to be addressed theoretically.

4.3.8.1 *Rebar.* All rebar is made of steel of one kind or another. While rebar use varies, it is probably never more than 5 % of the barrier area and much less on a volume basis, and therefore is a fairly small perturbation to the shielding. Steel has a much higher density than concrete (~7.8 versus 2.35 g cm^{-3}; or ~490 versus 147 lb $feet^{-3}$). The mass attenuation coefficient for photons is greater for iron than concrete below ~800 keV and above ~3 MeV.

Between these values, it is lower but not by enough to make up for the greater density of iron. Therefore, in all cases, if part of the concrete is replaced with steel, the result will be an improved shield for photons.

For neutrons, the situation is more complicated. Steel has no hydrogen and is quite transparent to neutrons in large area shields. On the other hand, it has quite a good scattering cross section (~3.5 barns) for both inelastic and elastic scattering in the region of a few million electron volts. Thus steel rebar will tend to produce longer path lengths through the concrete. Most of the rebar will have sufficient concrete behind it so that the use of removal cross sections is valid and that of iron is 0.16 cm^{-1} compared with 0.09 cm^{-1} for ordinary concrete (Rogers *et al.*, 1995; Walker and Grotenhuis, 1961). In addition, iron has quite a useful thermal neutron cross section of 2.55 barns. Therefore, it is expected that the addition of rebar will improve the neutron shielding that is calculated from pure concrete data.

4.3.8.2 *Form Ties.* Form ties completely penetrate the concrete shield. Traditionally, they are very heavy double wires or steel rods 2.5 cm (1 inch) or more in diameter that are cut to release the forms. There is now an option of using fiberglass form ties, which is discussed separately below. If appearance is at all important, the protruding ties are cut back below the concrete surface and grouted over to prevent rust from running down the concrete.

Such a form tie could be considered a steel-filled duct and could point directly toward the isocenter. The following considerations are relevant to the shielding effects of these form ties:

- Throughout the energy range of interest down to ~50 keV (only 3 to 4 % of the dose-equivalent results from neutrons <100 keV), the neutron cross section for steel is at least 3 barns with a removal cross section of ~0.25 cm^{-1}. This means that in 10 cm (~4 inches) roughly only 6 % of the neutrons will not undergo a scattering reaction that will put the neutron out of the steel into the concrete.
- Since the neutrons produced in an accelerator scatter several times before leaving the shielded head, the apparent source is quite large and such a small diameter duct will "see" only a portion of it.
- Since the streaming of neutrons through ducts scales quite closely with the square root of the cross-sectional area, a form tie acts like a very long duct.

- Since the neutron field penetrating through this duct will have an area of <4 or 5 cm^2 and the area of a neutron detector is several hundred cm^2, it will usually not be detected by standard methods.

Thus it seems doubtful that steel form ties are of any significant concern in accelerator shielding.

Fiberglass form ties are less common, to date. However, they may become more common in the future. They are rigid rods that poke through holes in the forms on each side with a reusable fastener slid over each end. Their composition is proprietary but they are apparently glass fibers loaded into a hard epoxy-like resin. After use, the forms are removed and the protruding rod is cut off with an abrasive disk, sanded flush and then can be painted over. One manufacturer's ties range up to 2.5 cm (1 inch) in diameter and have a measured density of 1.95 g cm^{-3}. With a composition of glass plus resin, the *TVL* for photons would not be much different than for concrete of the same density. Likewise, due to the high hydrogen content in the epoxy, fiberglass form ties should not create a neutron shielding problem. Notwithstanding these facts, the contractor should avoid having tie lines pointing directly at the isocenter from the primary barrier.

4.4 Joints, Concrete Slab Junctions

Protective barriers of solid block (or brick) construction *should* have mortar (without voids) of at least the same density as the block, and for multiple course construction the joints *should* be staggered. Joints between different kinds of protective material *should* be constructed so that the overall protection of the barrier is not impaired. Lead or steel barriers need to extend into adjacent concrete barriers in order to attenuate the radiation scattered in the concrete.

Openings in protective barriers for doors, windows, ventilating ducts, conduits, pipes, etc. may require radiation baffles to ensure that the required degree of overall protection is maintained. Whenever possible, the opening *should* be located in a secondary barrier where the required shielding thickness is less. The design of the baffles will depend upon a number of factors. These include:

- energy of radiation;
- orientation and field size of useful beam;
- size and location of opening in the protective barrier;

- geometrical relationship between radiation source and opening; and
- geometrical relationship between opening and persons, materials or instruments to be protected.

The methods of protection discussed in this Section are intended as guides only. Effective protection at minimum cost can be achieved only by consideration of all pertinent factors of each individual installation. Generally, the most economical shielding material for a baffle for photons is lead because the amount of radiation scattered from lead is less than that from lighter materials and the scattered radiation is more readily attenuated in lead. Since lead is not structurally self-supporting, it has to be mounted so that there will be no sagging.

Where ducts terminate at a grille in the wall surface of a secondary protective barrier, a lead-lined baffle may be required in front of the grille; the baffle needs to be at a sufficient distance to permit adequate flow of air and needs to extend far enough beyond the perimeter of the opening in order to provide the required degree of protection. This is illustrated in Figure 4.2.

Service boxes, conduits, etc. imbedded in concrete barriers may require lead shielding to compensate for the displaced concrete. For example, if the outside diameter of a steel conduit is large, if the conduit passes through the barrier in line with the useful beam, or if the concrete does not fit tightly around the conduit, compensatory shielding is required. Where supplementary lead shielding is required, its thickness *should* be at least equal to the

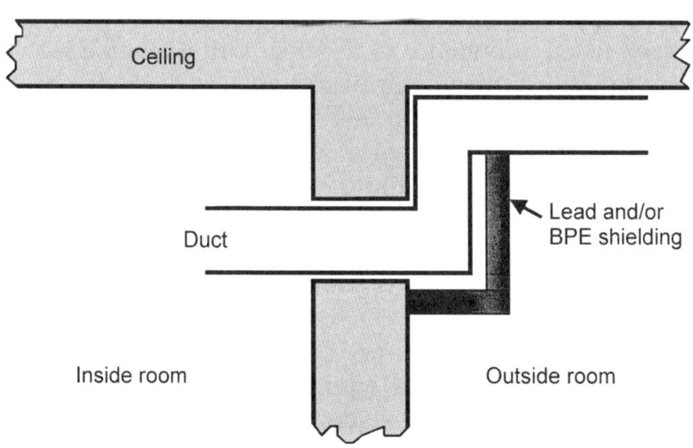

Fig. 4.2. Bend in duct to avoid radiation streaming.

lead equivalence of the displaced concrete. The lead equivalence of the concrete will depend upon the energy of the radiation and may be obtained from the ratio of the *TVL*s contained in Figures A.1a and A.1b (Appendix A). The shielding *should* cover not only the back of the service boxes, but also the sides, or extend sufficiently to offer equivalent protection.

4.5 Access to Radiation Vault

Access to radiation treatment rooms involves either a maze entry or a direct-shielded door. Such door shielding can be very heavy, even when located in a wall exposed only to leakage and scattered radiations. It may weigh several tons and require an expensive motor drive and will also require means for emergency manual operation (*e.g.*, during a power failure). Provision *should* be made to remove the patient by an emergency access in the event the heavy door cannot be opened even by manual methods. Automatic door systems *should* be provided with a mechanism to stop (or reverse) the direction of the motor drive if it collides with anything while it is opening or closing, and the operating mechanism *should* include an emergency stop control. Because of the complex critical nature of these systems, emergency door procedures *should* be posted and made part of the facility training. Also, there *should* be daily checks of the operation, and routine, periodic and detailed inspections of the structural and operational integrity of the total access system.

A maze arrangement generally is the most economical approach, as the shielding of the door can be greatly reduced, usually to <6 mm of lead, if it is exposed to multiple-scattered radiation only. It is therefore considered a part of the generic room design. The required lead equivalence of the door will depend upon radiation energy, maze design, weekly workload, and beam orientations. A typical maze design is shown in Figures 7.1 and 7.2 (Section 7). However, mazes do increase space requirements and provide less convenient access to treatment staff.

For higher-energy rooms, an extra door constructed of thermal neutron absorbing material and installed at the inside end of the maze usually will reduce the length of the maze needed or reduce the shielding requirements for the outside door.

Height and width of doorways, elevators and mazes *should* be adequate to permit delivery of equipment into treatment rooms, unless other provisions are made. Gamma-ray beam therapy installations also require access for replacement sources in large, heavy shipping containers.

While a lead door sill is often recommended (NCRP, 1976; 1977), lead is very transparent to neutrons and scattering under the door will be greater for neutrons than it is for x rays. Therefore, the lead sill is not recommended for the higher-energy accelerators (>10 MV).

Architects will often put 1.6 cm (5/8 inch) gyp board[15] on the outside of the concrete surfaces around doors for esthetic purposes. Gyp board has very low density and hydrogen content so it is a very poor neutron shield and its presence simply creates a larger crack for radiation leakage around the door. Depending upon how close to vertical and straight the wall turns out to be, there may or may not be a significant crack. If there is a significant crack, it *should* be filled with BPE or plain polyethylene. Polyethylene can even be used as a rubbing pad for the door to close on.

4.6 Ducts

Ducts can be divided into several types according to function and the related size. The largest ducts are usually for HVAC purposes, and two ducts (entry and return) are required for the treatment room. The cross-sectional dimensions of these ducts can be as large as 60 × 30 cm. The next largest ducts are usually for machine cables and these typically are 30 × 10 cm (cross-sectional dimensions). A circular duct, not <10 cm in diameter, is required for miscellaneous cables, such as those used for physics and QA purposes. This conduit, or one similar to it, may also provide access for real-time monitoring of patient specific parameters during treatments. Electrical and water ducts are circular in cross section and are typically <10 cm in diameter.

The purpose of correctly orienting the duct is to ensure that: (1) the least amount of concrete is displaced by the duct in the direction of the radiation beam (Figure 4.3), and (2) the direct radiation passing through the aperture is minimized. The ducts may exit the room at an angle to the wall to maintain this short path or they may be staggered through the wall. Ducts *should* never be placed in the primary barriers, no matter how small.

Specific requirements for different types of ducts are listed below in order of decreasing size of the duct.

[15]Gyp board is a common name for gypsum plaster board (a central layer of plaster with heavy paper glued on each side).

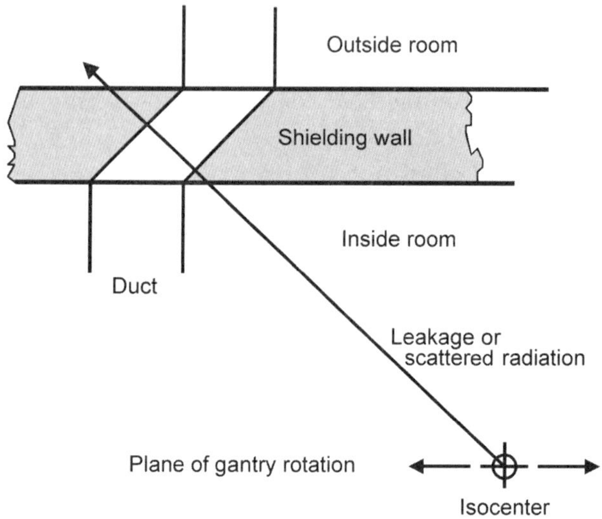

Fig. 4.3. Duct penetration with least amount of concrete removed along path of primary radiation.

4.6.1 *Heating, Ventilation and Air-Conditioning, and High-Voltage Ducts*

Because of their large cross-sectional area, it is important that HVAC and high-voltage ducts are placed in such a way that radiation passing through them will require the least amount of remedial shielding. This will depend on the highest energy available from the linear accelerator as well as the layout geometry. For the case where the ducts pass through the walls, it is important that the ducts are placed as high as possible to reduce the amount of downward scattered radiation and, hence, to minimize the exposure to personnel outside the room.

Three options are discussed below for: (1) rooms with mazes, (2) rooms without mazes, and (3) ducts that pass through the ceiling.

4.6.1.1 *Rooms with Mazes.* For rooms that incorporate a maze, the logical place for the duct penetrations is directly through the shielding above the door, where the photon and neutron fluences are lowest. To assess the need for additional shielding around the ducts, first assume that the photon and neutron dose-equivalent rate at the ducts are the same as those at the door. Then calculate the effect of this radiation scattered to a person directly outside the door. For low-energy machines (<10 MV) no additional shielding around the duct is generally required. For high-energy machines, McGinley (2002) has shown that, for an 18 MV accelerator, the need for additional shielding depends strongly on the length of the maze. For a maze 5.2 m in length, the total dose equivalent at the duct was found to be on the order 0.07 mSv week^{-1}, so no additional shielding was required. However, for a 2.2 m maze, the total dose equivalent was ~0.75 mSv week^{-1}, with roughly equal contributions from photons and neutrons when averaged over the four primary gantry directions. In this case, added shielding is needed and the preferred arrangement is to bend the ducts immediately after they have exited the maze (Figure 4.2). If that is not possible, the ducts *should* be wrapped with lead and BPE, as shown in Figure 4.4. McGinley (2002) reports that for an 18 MV accelerator, a dose-equivalent reduction of four for neutrons and two for photons will be produced by a 1.2 m long duct wrap composed of 2.5 cm BPE and 1 cm lead in a 3.6 m maze. For thermal neutrons, NCRP (1977) indicates that a factor of 10 can be gained by using a duct length of two to three times the square root of the duct cross section. A third alternative (albeit a more expensive option) is to use the concrete shielding as a baffle, as shown in Figure 4.5. Here two parallel, overlapping sections of concrete provide a vertical "mini-maze." For this arrangement to be successful, the degree of overlap *should* be as large as possible.

For rooms that include more than one bend in the maze, it is unnecessary to shield ducts that follow the lengths of the maze.

4.6.1.2 *Rooms without Mazes.* For rooms that do not have a maze, the walls parallel to the gantry rotation plane are best suited for duct penetrations, because the radiation shielding requirements are lower for these walls than for those in the gantry rotation plane. Since the whole length of the wall can be used for duct placement, the ducts can be angled as shown in Figures 4.2 and 4.3. Since the duct penetration is high up on the wall, 3.05 m (10 feet) above floor level, a simple calculation for 90 degree scattered radiation usually shows that the exposure from x rays to a person outside the wall will not exceed the shielding design goal if the direct shielding is adequate. However, the neutron dose equivalent outside a room from neutrons scattering through a duct is very difficult to calculate. As a practical measure, the tightest possible bend in the duct *should* be ensured, preferably outside the room, and the duct *should* be wrapped tightly with up to a 10 cm thickness of BPE.

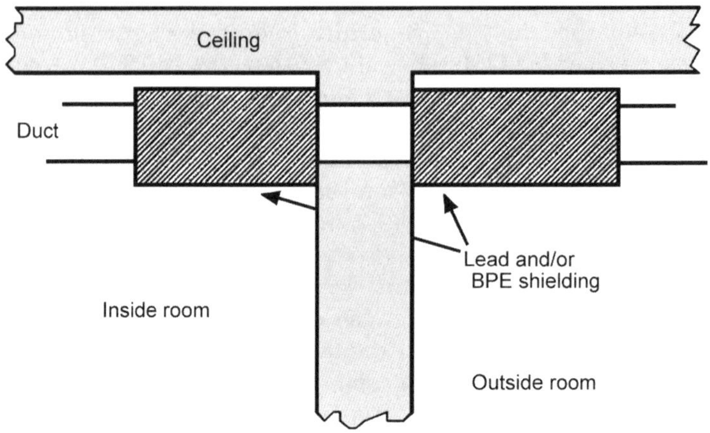

Fig. 4.4. Duct wrapped with shielding material on both sides of barrier.

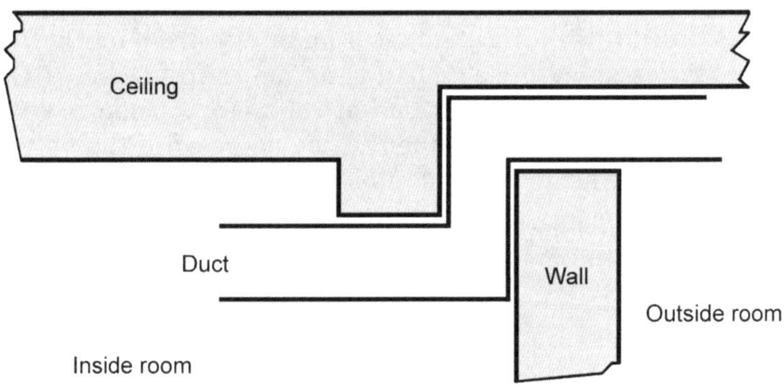

Fig. 4.5. Concrete baffle used to accommodate duct.

4.6.1.3 *Ducts Passing Through the Ceiling.* It is important to design a cross section for the duct that is rectangular with as high an aspect ratio as possible (*i.e.*, the highest ratio of the width to the height). Also, the secondary radiation from the target in the direction of the duct should be as orthogonal as possible to the axis of the duct and also to the longest side of the duct. If the duct is angled 90 degrees directly above the ceiling, appropriate shielding can be readily applied, if necessary, for occupied spaces above or below. However, if the extra distance to the floor above and the thickness of the floor above the ceiling (typically, 10 to 15 cm of concrete) is taken into account, it is possible that no extra shielding will be required.

4.6.2 *Machine Cables*

Machine cables are usually placed at floor level inside the room, often below the floor, and either angle up to the control area outside or pass directly outside, if below floor level. They generally do not require additional shielding, unless for some reason the console area is behind a primary barrier.

4.6.3 *Water and Electrical Conduits*

Water and electrical conduits are usually <2.5 cm in diameter and no special precautions are needed, provided the placement guidelines noted above are adhered to. It is unwise to build these pipes directly into the concrete form work because of possible failure and difficulty of replacement. Rather, a hole of slightly larger diameter than the required conduit is placed in the concrete form-work so that the conduit can be readily passed through it on installation. For conduits or pipes larger than 2.5 cm in diameter wrapping the pipe with lead should be considered in order to compensate for the missing concrete.

4.7 Lead-Only Rooms

When space is at a premium, it is natural to try to use only very high-density materials for shielding, since the thickness can be reduced by the ratio of the densities for photon shielding. Lead has some excellent properties in this respect with a density almost five times greater than that for concrete and it is relatively inexpensive and readily available. Lead is also easy to work with as bricks and can be cut to fit on location. Lead-only barriers are easy to demolish and hence they are again ideal for renovation work when limited space may have to be returned to use unshielded at a later date.

Among the chief drawbacks for lead is its lack of structural integrity and thus the need to be held in place by steel or concrete. Also, it is a known toxin which requires special provisions during cutting or machining. The most important drawback, however, in the use of lead for shielding of linear accelerators is its high neutron-production cross section for high-energy photons and its low-absorption cross section for neutrons. These two factors restrict the use of lead for accelerators operating above 10 MV. At higher energies, the neutrons from the accelerator are not absorbed significantly by the lead and additional neutrons are created in the lead by the high-energy photons. Thus there needs to be some added shielding for these neutrons beyond the lead barrier, even if the barrier is adequately thick to stop the photons. Typically, this will require up to 0.91 m (3 feet) of concrete or BPE equivalent in addition to the lead in a primary barrier.

For x rays of 10 MV or less, lead can be used for shielding with certain precautions. As stated earlier, structural engineering calculations are needed to ensure that the lead is held in place by steel or some other support structure. If bolts are used, they need to be anchored deeply enough to hold the lead weight and care needs to be taken that the bolts cannot pull through the malleable lead. Properly anchored steel bolts are usually of such length that they attenuate the photons at least as well as the length of lead through which they penetrate. Brick or block lead needs to be mated so as to eliminate any streaming of radiation through joint gaps. This is often accomplished by means of bricks which are fabricated with tongue and groove edges.

Penetrations through the lead (such as ducts) are easy to cut on site but special care needs to be taken since the dimension of the opening relative to the width of the barrier determines the absorption of x rays that are diagonally incident on the barrier. Hence a 7.6 cm (3 inch) diameter opening through 0.91 m (3 feet) of concrete attenuates x rays that are diagonally incident more effectively than does the same 7.6 cm (3 inch) opening through 15.2 cm (6 inches) of lead. Likewise conduits running through lead are a more significant void because each 2.5 cm (1 inch) of missing lead is equivalent to 15.2 cm (6 inches) or more of concrete.

Groundshine radiation may be a problem, especially for higher-energy x rays, if the treatment room wall is constructed of metal and polyethylene. As can be seen from the Figure 4.6, little shielding is provided by the concrete floor slab when the beam is aimed at the junction between the wall and the floor. To rectify this problem it will be necessary to add steel or lead to the floor (as shown by the dashed lines) in order to reduce the scattering path length under the wall. Alternatively, the lead and polyethylene wall can be extended into the floor.

4.8 Beamstoppers

Though it is becoming rare, some linear accelerators are installed with either a fixed or a retractable beamstopper. The purpose of the beamstopper is to attenuate the primary beam by a factor of 10^3 so that the walls and ceiling that would otherwise be primary barriers become secondary barriers, thus saving on construction costs and space. This attenuation is achieved by encasing a sufficient thickness of lead in a steel case having a diameter that will intercept the largest primary beam with a margin of 10 cm and the beamstopper is usually located at a distance of ~1.5 m from the target. This diameter is usually sufficient to intercept scattered

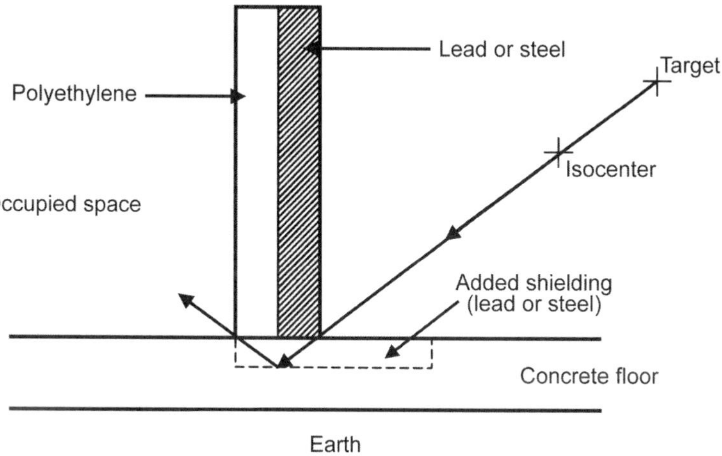

Fig. 4.6. Possible groundshine radiation may require added shielding in the area bordered by the dashed lines.

radiation with angular incidence up to 30 degrees. The disadvantage of a linear accelerator having a beamstopper is clearance with the patient support assembly as well as real-time beam imaging devices and, perhaps more importantly, the ability of the therapist to move around the couch to set up and verify the correct positioning of the patient.

Beamstoppers were used very frequently with ^{60}Co teletherapy units and the practice was often continued when the units were replaced with linear accelerators to avoid the necessity of making any shielding changes. However, the concept of laminated shielding (Section 2.2.3) where either lead or steel is added to existing concrete walls came into use and the disadvantages of using beamstoppers noted above severely outweighed the one time cost of renovating the shielding.

5. Special Considerations

5.1 Skyshine Radiations

Some radiation facilities are designed with little shielding in the ceiling above the accelerator. A problem may then arise as a result of the radiation scattered by the atmosphere to points at ground level outside the treatment room. Stray radiation of this type is referred to as skyshine, and NCRP (1977) gives methods for the calculation of skyshine for accelerator facilities. This situation is illustrated in Figure 5.1 for photon skyshine. Following the method of McGinley (2002) and using the terms shown in Figure 5.1, the dose-equivalent rate [$\dot{H}$ (nSv h^{-1})] at a distance d_s (meters) from the isocenter is given by Equation 5.1.

$$\dot{H} = \frac{2.5 \times 10^{7}(B_{\text{XS}}\, \dot{D}_{\text{O}}\, \Omega^{1.3})}{(d_{\text{i}}\, d_{\text{s}})^{2}} \tag{5.1}$$

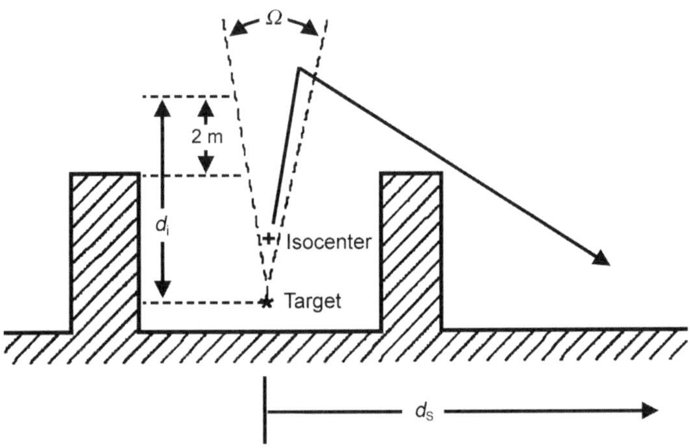

Fig. 5.1. Simplified schematic showing the accelerator isocenter and target location for photon skyshine evaluation.

In Equation 5.1:

B_{xs} = roof shielding transmission factor for photons
Ω = solid angle of the maximum beam (steradians)
d_i = vertical distance from the target to a point 2 m above the roof (meters)
$\dot{D}_0$ = x-ray absorbed-dose output rate at 1 m from the target (Gy h^{-1})

and the constant 2.5×10^7 includes a conversion of gray to nanosievert (nSv).

Similarly, neutron skyshine radiation is also covered in NCRP (1977). The geometry for this component is shown in Figure 5.2 where the solid angle is now determined by the walls of the facility rather than the collimator on the treatment unit. Note that, in the case of neutrons, the x-ray beam is considered to be pointing downward so the target is at its highest point. The dose-equivalent rate is given by Equation 5.2.

$$\dot{H}_n = \frac{0.85 \times 10^5 \, H_{ns} \, \dot{\Phi}_0 \, \Omega}{d_i^2} \quad (\text{for } d_s \leq 20 \text{ m}) \tag{5.2}$$

In Equation 5.2:

$\dot{H}_n$ = neutron dose-equivalent rate at ground level (nSv h^{-1})
H_{ns} = ratio of the dose equivalent 2 m beyond the ceiling shield to the neutron fluence incident at the ceiling (Sv cm^2 n^{-1}) [values for H_{ns} can be obtained from Figure A.2 as adapted from NCRP (1977)]

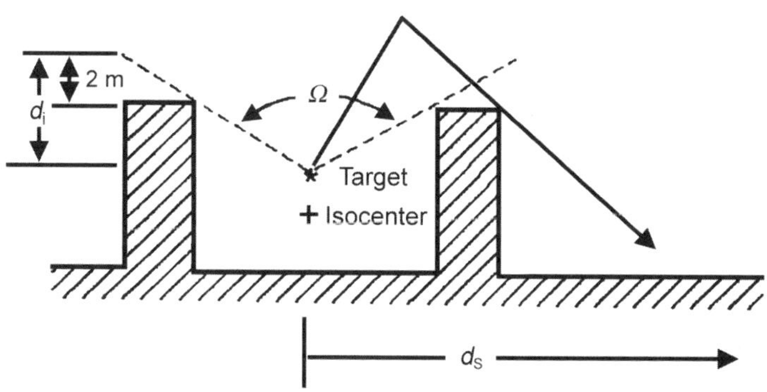

Fig. 5.2. Irradiation conditions for the neutron skyshine evaluation.

d_i = distance from the target to the ceiling plus 2 m (meters)
$\dot{\Phi}_o$ = neutron-fluence rate at 1 m from the target (n cm^{-2} h^{-1})
Ω = solid angle of the shield walls subtended by the target

and the constant 0.85×10^5 includes a conversion from sievert to nanosievert.

Curves of dose equivalent for monoenergetic, unidirectional broad neutron beams perpendicularly incident on concrete barriers are shown in Figure A.2 as functions of shielding thickness (g cm^{-2}). The quantity dose equivalent contains the fluence-to-dose-equivalent conversion (including the gamma-ray contribution), based on the spectrum at the shielding depth in question. The intercept at 0 g cm^{-2} is equivalent to the fluence-to-dose equivalent conversion coefficient for no shielding (NCRP, 2003).

Note that the fast neutron-fluence rate can be calculated from the Q_n values in Table B.9 (Appendix B) and the first term of Equation 2.16 that represents the direct component of the photoneutrons from the accelerator using Equations 5.3, 5.4a, and 5.4b.

$$\dot{\Phi}_o = \varphi_{dir}\,\dot{D}_o \tag{5.3}$$

$$\dot{\Phi}_o = \left[\frac{\beta Q_n}{4\pi(100)^2}\right]\dot{D}_o \tag{5.4a}$$

or

$$\dot{\Phi}_o = \beta\,Q_n\,\dot{D}_o\,(7.96 \times 10^{-6}) \tag{5.4b}$$

In these equations, $\dot{D}_o$ is the x-ray absorbed-dose rate at 1 m (Gy h^{-1}).

McGinley (1993) has compared skyshine measurements made at an 18 MV medical accelerator facility (having a roof that provided only 10 % attenuation of the x-ray beam) with values calculated using the techniques presented in NCRP Report No. 51 (NCRP, 1977). The McGinley (1993) results are given in Table 5.1 and Table 5.2 for photons and neutrons, respectively. As can be seen, there is very poor agreement between the calculated and measured values. Thus, Equations 5.1 through 5.4 should be used with caution to get (at best) an order of magnitude estimate of these phenomena and to indicate which room design parameters will have an impact on the resultant skyshine. The measured photon data show an increase as with distance away from the barrier until a maximum is reached at a distance which is about equal to the height of the barrier, beyond which there is then a slow (almost linear) fall off with distance.

TABLE 5.1—*Measured and calculated x-ray skyshine dose-equivalent rates for 18 MV x rays produced in a vault with little ceiling shielding (adapted from McGinley, 1993).*[a]

Distance from Isocenter (d_s) (meters)	Photon Dose-Equivalent Rate (μSv h^{-1})		Ratio Measured / Calculated
	Measured	Calculated	
7.5	50	202	0.25
9.4	112	127	0.88
10.6	150	97	1.5
13.6	157	63	2.5
19.2	100	30	3.3
25.4	75	18	4.2
33.0	55	10	5.3
48.3	25	4.7	5.3

[a]The absorbed-dose rate at isocenter was 4 Gy min^{-1} (240 Gy h^{-1}) with the beam directed vertically up; d_i = 5.97 m and Ω = 0.122 in Equation 5.1.

TABLE 5.2—*Measured neutron skyshine for an 18 MV accelerator with little ceiling shielding and an absorbed-dose rate of 240 Gy h^{-1} at 1 m (McGinley, 1993).*

Distance from Isocenter (d_s) (meters)	Neutron Dose-Equivalent Rate (μSv h^{-1})	
	Measured	Calculated from Equation 5.2
5.14	68	8
8.5	209	8
11.2	187	8
14.3	151	8
17.3	130	8
18.9	104	8
20.8	83	NA[a]

[a]NA = not applicable.

The caution required for the use of this method is shown by the fact that Equation 5.2 gives a computed value[16] of ~8 μSv h^{-1} for the neutron skyshine (without any shielding) for all distances out to 20 m, which is 10 to 30 times less than the measured values.

As stated in NCRP Report No. 144 (NCRP, 2003), "These simple recipes must be treated with great caution, particularly so when extrapolated to large and small distances." Notwithstanding this uncertainty, it is instructive to consider such calculations since they demonstrate the rough order of magnitude of the skyshine component relative to other sources such as barrier-transmitted or background radiation, and the calculations indicate the empirical factors that affect the skyshine contribution.

5.2 Side-Scattered Photon Radiation

In addition to skyshine, Zavgorodni (2001) has presented data demonstrating the significance of x rays scattered laterally from thin roof barriers to adjacent structures as shown in Figure 5.3. This work does not treat either oblique incidence of the photon beam on the ceiling or the production of photoneutrons in the roof; yet, for the cases considered, the scattered photons predominate

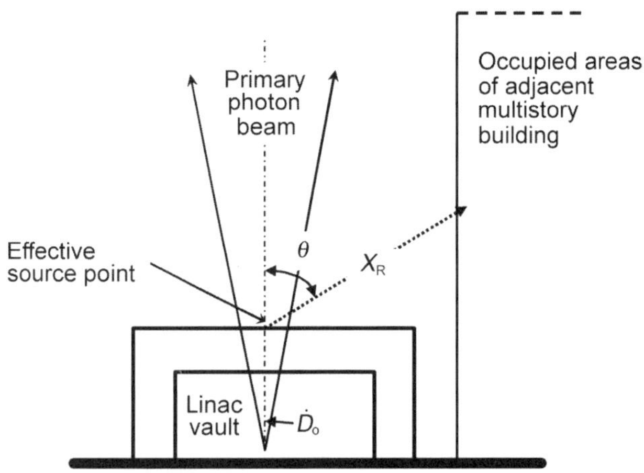

Fig. 5.3. Side-scattered photon radiation from ceiling barrier (adapted from Zavgorodni, 2001).

[16] Q_n = ~10^{12} neutrons Gy^{-1}; $\dot{D}_O$ = 240 Gy h^{-1}; β = 1 (Equation 2.16); Ω = 2.7 steradians; H_{ns} = 4 × 10^{-10} Sv cm^2 for zero thickness; E_n = 1.5 MeV [from Figure A.2, adapted from NCRP (1977)]; and d_i = 4.7 m.

over both leakage radiation and skyshine. The measurements and calculations covered multiple geometries and energies and Monte-Carlo calculations were used to treat the angular distribution of the divergent photon beams transmitted through the concrete roof. Equation 5.5 was derived empirically for the side-scattered radiation contribution.

$$\dot{H}_{SS} = \frac{\dot{D}_0 \, F \, f(\theta)}{d_R^2 \, 10^{\left[1 + \frac{(t - TVL_1)}{TVL_e}\right]}} \tag{5.5}$$

In Equation 5.5:

$\dot{H}_{SS}$ = side-scattered dose-equivalent rate (Sv h^{-1})
$\dot{D}_0$ = x-ray absorbed-dose output rate of the accelerator at 1 m from the target (Gy h^{-1})
F = area of the square field (m^2) at 1 m from the target
$f(\theta)$ = angular distribution of the roof-scattered photons from Table 5.3
X_R = distance from the beam center at the roof-top to the point of interest
t = roof thickness (meters)
TVL_1 and TVL_e = first and equilibrium tenth-value layers (meters) of the primary radiation in the roof shielding material, respectively (Table B.2)

After using Monte-Carlo methods to compute the angular distributions of the roof-scattered radiations for energies of 4, 6, 18 and 23 MV, the resulting distributions were fit with a polynomial that was largely independent of the beam energy for angles up to 85 degrees. Values for use in Equation 5.5 are given in Table 5.3.

5.3 Groundshine Radiation

Groundshine radiation may be a problem when a high-energy treatment room is constructed of thin laminated barriers. As can be seen from the Figure 4.6, little shielding is provided for the penetrating x rays by the concrete floor slab when the beam is aimed at the junction between the floor and a primary wall composed of lead and polyethylene. Neutrons will be attenuated adequately by the concrete in the floor and need not be considered. To rectify this problem a lead slab (dashed lines on Figure 4.6) can be added to the floor of the occupied space or the lead and polyethylene wall can be extended into the floor. Barriers composed of thin concrete-lead laminates can also exhibit this problem.

TABLE 5.3—*Angular distribution function f(θ) for photons side-scattered from the roof (values interpolated from Zavgorodni, 2001).*

Angle (θ)	Angular Distribution Function [$f(\theta)$]
20	0.38
30	0.26
40	0.16
50	0.10
60	0.065
70	0.035
80	0.014
85	0.005

5.4 Activation

High-energy accelerators (>10 MV), expose both patients and personnel to radionuclides created by neutron and gamma-ray activation of materials within the treatment room. While this issue has been studied by many authors, (Almen *et al.*, 1991; LaRiviere, 1985; McGinley *et al.*, 1984; O'Brien *et al.*, 1985), the most complete work to date is that of Rawlinson *et al.* (2002). Rawlinson *et al.* (2002) indicates that at 18 MV the principal radionuclides produced result from (n,γ) reactions (Table 5.4), and the report attributes the sources for these radionuclides to: the aluminum (^{28}Al) in the couch frame, antimony (^{122}Sb) in the lead shielding in the gantry head of the accelerator, and multiple sources throughout the room (^{56}Mn and ^{24}Na).

TABLE 5.4—*Principal radionuclides identified at 1 m lateral to isocenter of Varian Clinac 21EX Linear Accelerator (Rawlinson et al., 2002).*

Radionuclide	Half-Life	Probable Nuclear Reaction	Decay Mode	Principal Gamma-Ray Energies (MeV)
Al-28	2.3 m	^{27}Al(n,γ)^{28}Al	β^-, γ	1.78
Mn-56	2.6 h	^{55}Mn(n,γ)^{56}Mn	β^-, γ	0.85, 1.81, 2.11
Na-24	15.0 h	^{23}Na(n,γ)^{24}Na	β^-, γ	1.37, 2.75
Sb-122	2.8 d	^{121}Sb(n,γ)^{122}Sb	β^-, β^+, γ	0.51, 0.56

Rawlinson *et al.* (2002) determined that after 48 h the dose-equivalent rates are very close to the background level in the room prior to irradiation and hence there is no appreciable buildup of activity over the long term. Also, since the dominant radionuclides are produced by thermal neutrons, with fluence proportional to the fast neutron fluence, the activation dose equivalents scale with the neutron leakage radiation and remain rather insensitive to the radiation beam energy due to the limited range of medical electron accelerator energies (NCRP, 1984).

Table 5.5 summarizes the predicted activation deep dose equivalents (see Glossary) to treatment staff for different treatment regimes (Rawlinson *et al.*, 2002). Comparison of the Rawlinson *et al.* (2002) results to those of other published efforts indicates very good agreement, as summarized in Table 5.6.

It is noteworthy that current IMRT treatments result in lower cumulative deep dose equivalents to staff due to the large amount of time necessary for both patient and machine setup. However, it is expected that with increased efficiencies in these areas, there will be an increase in the number of patients treated, and therefore an increase in the deep dose equivalents to the staff. There is little in the way of shielding design that will affect the deep dose equivalent to staff received from activation radionuclides, however Rawlinson *et al.* (2002) make the following specific recommendations in order to ensure compliance with international effective dose limits:

- IMRT treatments *should* be delivered at low x-ray energies;
- manufacturers *should* design accelerators to minimize neutron production and avoid the use of aluminum and other materials which have a high neutron-capture cross section; and
- irradiations involving high-energy x rays, especially physics and QA measurements, *should* be scheduled at the end of the day in order to allow for overnight decay of the longer-lived radionuclides.

5.5 Ozone Production

Although it is not related directly to radiation shielding concerns, the production of ozone (O_3) by electron accelerators does represent a safety hazard and it was treated in Appendix I of NCRP Report No. 51 (1977). NCRP (1977) recommends that the concentration of ozone *should* not exceed 0.1 ppm for continuous exposure

TABLE 5.5—*Activation deep dose equivalents (to staff) for different treatment regimes (Rawlinson et al., 2002).*

Treatment Regime		Weekly Deep Dose Equivalent (μSv week^{-1})					Annual Deep Dose Equivalent (mSv y^{-1})
		^{28}Al	^{56}Mn	^{24}Na	Long-Lived	Total	
Conventional radiation therapy	Benchmark treatment regime	26.2	23.2	5.9	5.6	60.9	3.2
	Four field	21.6	22.2	5.8	5.6	55.2	2.9
	Mixed energies	13.0 H[a] 0.1 L[b]	6.2 H 5.7 L	1.5 H 1.5 L	1.4 H 1.4 L	30.8	1.6
	Benchmark treatment regime, with QA[c]	27.9	31.4	7.9	7.2	74.2	3.9
Intensity-modulated radiation therapy	Current IMRT	6.1	17.1	4.3	4.0	31.5	1.6
	High-efficiency IMRT	50.5	59.9	15.1	14.4	139.9	7.3
	Future IMRT	108.3	99.7	25.1	24.0	257.1	13.4

[a]H = "high" energy used.
[b]L = "low" energy used.
[c]QA = quality assurance use of the accelerator.

TABLE 5.6—*Deep dose equivalent to therapist from high-energy activation in treatment room.*

Accelerating Voltages	Microsievert per Patient	Millisievert per Million Monitor Units	Reference
18	0.4	1	Rawlinson *et al.* (2002)
13 to 17	0.2		Almen *et al.* (1991)
24	0.18		LaRiviere (1985)
18	0.29		LaRiviere (1985)
25		1.2	O'Brien *et al.* (1985)

and it was observed that electron beams are much more efficient producers of ozone than photon beams since it is the electron interaction with the oxygen molecule which produces the ozone. McGinley (2002) employed the equations of NCRP (1977) and concluded that "for normal clinical use of electron beams, a room ventilation rate of about three room changes per hour is more than adequate for health protection."

5.6 Tomotherapy

One of the newest methods of radiation therapy treatment involves the combination of the principles of computed-tomography imaging with IMRT. A linear accelerator waveguide, producing about a 6 MV accelerating voltage, is mounted in a computed-tomography-type gantry with a multileaf collimator and a fan-like beam that is a few centimeters long by ~40 cm wide. This arrangement allows the unit to be rotated around the patient while the radiation beam intensity is modulated and the patient is translated through the gantry opening. With computer optimization and intra-treatment imaging of the daily patient setup anatomy, very precise absorbed-dose treatment distributions may be achieved.

The net result is that the primary beam, though reduced in width and intercepted by a beamstop opposite the patient, rotates around the patient many times due to the required indexing of the narrow beam as well as the attenuation of much of the beam in order to achieve the modulated intensity. Thus, there is a very large C_I value (Section 3.2.2) because there are many more monitor units than centigrays delivered to the isocenter.

Each of these factors was studied by Robinson *et al.* (2000) with the following conclusions:

- primary shielding barriers for tomotherapy may be up to 10 times narrower than that required by a conventional field size, while their thickness may be up to 1 *TVL* greater;
- secondary shielding barrier requirements for scattered radiation are about the same for both tomotherapy and conventional methods because the integral absorbed dose delivered to the patient remains the same; and
- secondary shielding barrier requirements for the leakage-radiation component may need to be at least 2 *TVL*s thicker than is required for conventional treatments as a result of the modulation factors and indexing needed to treat the whole length of the tumor volume.

Thus, retrofitting an existing facility could be problematic if all of the secondary barriers require the addition of 2 *TVL*s, while the primary barrier requires one added *TVL* over its central portion. These changes may be offset, however, if the leakage radiation from the accelerator head is reduced by a *TVL* or more and the beamstop on the beam exit side of the patient is 2 to 3 *TVL* equivalents greater than considered in the preliminary design specifications as used by Robinson *et al.* (2000). Therefore, careful note must be taken of the actual design of the unit that is to be installed, and a distinction needs to be made between serial tomotherapy, as used by Robinson *et al.* (2000) with the patient indexed through the beam, and helical tomotherapy where the patient moves continuously. It is likely that these conclusions do not apply to helical tomotherapy since it employs greatly increased internal shielding and makes more efficient use of the rotating beam.

5.7 Robotic Arm

A SRS device consisting of a 6 MV x-ray source mounted on a robotic manipulator can, in principle, point the primary beam at all barrier walls (Rodgers, 2005). For the CyberKnife® SRS machine made by Accuray, Inc. (Sunnyvale, California), the average ratio of monitor units to centigray delivered leads to a C_I of ~15 (Section 3.2.2). The largest CyberKnife® field size at the standard treatment distance (80 cm) is 6 cm in diameter and therefore the fields incident on barriers are relatively small compared to conventional treatment procedures. The recommended workload per treatment session is 12.5 Gy at the nominal treatment distance of 80 cm from the x-ray target (or 8 Gy at 1 m). This workload depends on the mix of single-fraction SRS to fractionated-SRS treatments per week. If there are 20 stereotactic treatment sessions (five being single fractions) per week, the weekly workload is 250 Gy at 80 cm (or 160 Gy week^{-1} at 1 m) and the leakage-radiation workload is 3.8×10^5 MU week^{-1}. For the primary beam, the recommended use factor is 0.05. It is noteworthy that based on the recommended values for C_I and U, the primary-beam contribution is approximately a factor of three higher than the leakage-radiation contribution.

Practical and safety limitations (such as treatment head clearance beneath the current couch model) keep the area of the ceiling irradiated by the primary beam small. Details of the range of solid angles over which the beam can be pointed *should* be examined with current data from the manufacturer in preparing a shielding plan. Section 7.2 contains an example shielding design calculation for a robotic arm type treatment unit.

5.8 Dedicated Intraoperative-Radiotherapy Units

Linear accelerators that produce only electron beams are used within operating suites in which direct access to the tumor can be achieved. Shielding assessment of such a mobile electron accelerator was considered by Daves and Mills (2001) and they found that these IORT units could be used in standard operating rooms without added shielding if the machine on-time is restricted to ~30 min week^{-1}. This results from: (1) the very low beam currents used for electrons only, (2) the low leakage radiation because no bending-magnets are employed, (3) the low bremsstrahlung production from the low-Z materials in the beam path, (4) the use of a compact beamstop beyond the tumor volume, and (5) low energy to eliminate neutron production. Each of these conditions needs to be carefully assessed to verify that the condition is applicable for the equipment under consideration. Adequacy of the radiation shielding design relies on detailed measurements of the operating unit in the actual working environment.

5.9 Cobalt-60 Units

Cobalt-60 teletherapy units house a ^{60}Co source with an activity of up to 10^{16} Bq and thus they present a potential radiation hazard even when there is no electrical power supplied to the unit. Provisions need to be made to ensure that the source always returns to the "safe" housed position at the completion of each irradiation and the source status is always indicated visibly both within the room and at the control console. During beam-on time, the shielding techniques for low-energy treatment machines covered in this Report are applicable to these gantry-mounted ^{60}Co teletherapy units. Most commonly, the treatment rooms employ a maze to avoid the use of heavy doors and the teletherapy unit is mounted to allow rotation around the isocenter. Other units do not use isocentric mounting but are capable of source head movements up and down. Some units are capable of head-swivel motions to direct the beam at an angle away from the isocenter. Although source-to-isocenter distance is traditionally 0.8 m, units with 1 m source-to-isocenter distance are available. The workload (W) for the 80 cm units is, therefore, different from the weekly absorbed dose at the isocenter as W is defined at a point 1 m from the source. When a beam interceptor is provided, it *should* transmit not more than 0.1 % of the useful beam. It will also reduce, by the same factor, the radiation scattered by the patient along a path shielded by the beam interceptor. For models in which the source housing can swivel the center of the useful beam away from the center of the beam interceptor, additional structural shielding is usually required. Electrical or mechanical means *shall* be provided to prevent irradiation when the useful beam is directed toward a barrier that is not a primary barrier.

The barrier thickness requirements may be calculated using methods described in Section 2. Barrier thicknesses thus obtained usually are adequate to meet the NRC requirement of 0.02 mSv in-any-one-hour (NRC, 2005a), except when the workload is unusually low. After the barrier thickness is obtained, the attenuated instantaneous dose-equivalent rate *should* be estimated at the point of interest assuming the maximum absorbed-dose output rate. Equation 3.14 may be applied to estimate the TADR in-any-one-hour (R_h), using the weekly workload and the estimated maximum number of normal patient procedures that could be delivered "in-any-one-hour" as described in Section 3.3.2. If the TADR exceeds the "in-any-one-hour" limit, additional shielding *shall* be specified. In other words, the total shielding must be adequate to reduce the radiation level to values below the weekly shielding design goal (Section 1.4), as well as the NRC "in-any-one-hour" limit on the TADR.

6. Shielding Evaluation (Surveys)

New facilities and old installations that have been modified to accept a higher energy accelerator or new types of procedures *shall* have a radiation shielding evaluation including a survey performed. The qualified expert performing the evaluation *shall* meet the standards set forth by the state radiation control agency. In general, the qualifications required are similar to those needed for certification as a medical or health physicist.

The radiation shielding evaluation *shall* include a review of the calculations that were used to determine the shielding specifications, an inspection during construction, measurements outside each barrier after the accelerator is installed, and an evaluation of the adequacy of the shielding. All occupied areas near a radiation installation *shall* be evaluated for the purpose of determining whether the shielding design goal is achieved and whether any person is likely to receive more than the recommended applicable annual effective dose values for controlled and uncontrolled areas (Section 1.4). A survey report *shall* be prepared and copies provided for the state, facility manager, and therapy physicist who is responsible for the accelerator. A copy of the report *shall* remain on file at the treatment facility. Approval or disapproval *should* be judged on the basis of compliance with the applicable NCRP recommendations and pertinent federal, state and local regulations. These are often derived from suggested state regulations produced by the Conference of Radiation Control Program Directors and elucidated in their documents (CRCPD, 1991; 1999). And, in the case of ^{60}Co machines, they are derived from the regulations of NRC (2005b; 2005c). If modification of the shielding is required, then a follow up survey *shall* be carried out and a revised survey report prepared.

6.1 Construction Inspection

An inspection of the facility during the construction phase is recommended. The inspection *should* include an evaluation of the following items.

- the location and width of the primary barriers relative to the proposed isocenter;
- the thickness and density of concrete as well as thickness and type of any other material used in the barriers;
- the thickness of metal shielding and polyethylene used for neutron shielding;
- the adequacy of direct-shielded doors, which require special inspection since they leave little room for error. That is:
  - make sure the concrete wall at the door is plumb over vertical within 3.2 mm (1/8 inch)
  - check that special shielding materials (*e.g.*, lead, polyethylene) are in place before the door is put up
  - measure door overlap at top and sides of door
  - measure door gap at top, sides, and bottom before door finishing materials are installed. It is almost impossible to make these measurements later
- the thickness of metal behind recesses in the concrete (*e.g.*, laser boxes);
- the thickness and composition of the HVAC shielding baffle if one is used;
- the location and size of conduit or pipe used for physics cables; and
- verification that the shielding design has been followed (*i.e.*, comparison to final shielding specification report).

More than one visit to the construction site may be needed in order to perform all of the tasks adequately.

A letter outlining the results of the construction inspection *shall* be prepared by the qualified expert and forwarded to the owner of the facility and the architectural firm involved in the construction. Any items of noncompliance *should* be clearly indicated and recommendations for corrections *should* be made.

6.2 Interlocks, Restrictive Devices, and Radiation Warning Lights and Signs

The testing of safety devices such as door interlock switches, limit switches for beam orientation, mechanical stops, etc. *shall* be performed after the installation is completed. These devices *shall* also be checked periodically.

The testing of the beam "on/off" control mechanism *shall* include a demonstration that with the beam in the "on" condition:

- the action of opening the door to the radiation therapy room interrupts the beam "on" status causing the useful beam to go to the "off" condition. For further information see NCRP Report No. 102 (NCRP, 1989); and
- the beam does not turn "on" again when the interlock circuit is restored until the equipment is manually activated from the control console.

The presence of appropriate warning signs and lights *shall* be determined. A red warning signal light (energized only when the useful beam is "on") *shall* be located: (1) in the control area and (2) near the entrance(s) to megavoltage or gamma-ray beam therapy rooms in addition to other appropriate locations in the treatment room (NCRP, 1989).

Emergency action procedures for gamma-ray beam therapy installations *shall* be posted near the control panel. "Radiation area" (NRC, 2005d; 2005e; 2005f) warning signs *should* be posted in all areas accessible to individuals in which radiation levels could result in an individual receiving a deep dose equivalent in excess of 0.05 mSv (5 mrem) in-any-one-hour at 30 cm from the radiation source or from any surface that the radiation penetrates. Appropriate "high radiation area" warning signs *shall* be posted at the entrance to any area wherein an individual could receive a deep dose equivalent from external sources in excess of 1 mSv (100 mrem) in-any-one-hour at 30 cm from the source or penetrated barrier (NRC, 2005d; 2005e; 2005f). Signs for "very high radiation area" are appropriate when the absorbed dose that an individual could receive in-any-one-hour from external sources would be in excess of 5 Gy (500 rad) at 1 m from the radiation source or 1 m from any surface that the radiation penetrates (NRC, 2005d; 2005e; 2005f).

Exceptions to the posting requirements may be permitted in a hospital or clinic room used for external beam therapy provided entrance to the area is strictly controlled and personnel in attendance take necessary precautions to prevent the inadvertent exposure of workers, other patients, and members of the public to radiation in excess of the posting limits. The qualified expert *should* check for local variations on this NRC regulation.

6.3 Radiation Survey

Immediately after the accelerator has been made operational, a preliminary survey *shall* be carried out to ensure that radiation exposures to the installation engineer and personnel near the facility do not exceed the applicable shielding design goal and time averaged dose-equivalent rate. Once the accelerator has been made completely operational and an initial calibration has been completed, a complete radiation survey *shall* be conducted.

One of the first things to be evaluated is the head-leakage radiation of the unit, since, if this is higher than the design specification, other survey results will also be higher. The leakage-radiation hot spots may be located with film wrapped around the head of the unit and then integrated readings can be made with an integrating type survey meter at appropriate distances from the head of the unit and in the patient plane.

Voids, cracks or other defects in the shielding are located using a sensitive photon rate meter with a fast response time, such as a Geiger-Mueller, scintillation or other sensitive rate meter. The maximum absorbed-dose output rate and largest field size are used. Any detected hot spots *shall* be followed up with a dose equivalent per monitor unit assessment. A calibrated ion chamber or other instrument having small energy dependence *should* be used to determine the exposure rate or integrated dose equivalent (H) in the areas being surveyed.

Once the barriers have been scanned to locate radiation hot spots, measurements are made outside of each barrier using a portable ionization chamber. This survey instrument *should* have both rate and integrate modes and adequate range(s) to cover expected measurements (*e.g.*, up to 50 mGy h^{-1}). For accelerators operating above 10 MV, a portable neutron survey instrument *should* be used to determine the neutron dose equivalent (H) per monitor unit (MU) and the dose-equivalent rate ($\dot{H}$). Each of the instruments used for the final measurements *shall* have a current calibration traceable to the National Institute of Standards and Technology. Integrated H measurements are normalized to the monitor unit, while $\dot{H}$ measurements are normalized to the MU min^{-1}.

The primary barriers are surveyed utilizing maximum field size without a phantom in the beam. Gantry angles of 0, 90, 180 and 270 degrees, as well as oblique angles that intercept the wall-ceiling intersections are commonly used. Wall-floor intersections are of concern where the thickness of the wall might allow for groundshine (Section 5.3). These measurements are made at 30 cm from each barrier surface. Other primary barriers are checked based upon the unique concerns of the final room construction. Radiation levels at further distances from the primary barrier can be estimated in a conservatively safe manner by means of the inverse square law as measured from the isocenter (McGinley, 2001b)

The secondary barriers are surveyed with a phantom in the beam and the collimator fully opened. If the region outside a barrier is a controlled or low occupancy area, the survey should also include any of the areas that are just beyond the controlled or low-occupancy area.

Measurements *should* also be made outside the facility to determine if radiation skyshine is present. When a tall building is located adjacent to the treatment facility measurements *should* be made on the adjacent upper floors to determine if skyshine, or side-scattered radiations can be detected.

If it is decided that an area requires on-going evaluation after the facility begins operation, either film, optically-stimulated luminescence dosimeters, or thermoluminescent dosimeters (TLDs) can be used for photons, while CR-39® dosimeters (PPG Industries, Inc., Pittsburgh, Pennsylvania), bubble detectors, or a hybrid TLD can be used for neutrons. Proper account needs to be taken of the detector energy dependence and its reciprocity (if data are to be collected over long periods of time). Also, careful placement of these detectors is required when monitoring the primary radiation because these fields can be highly position sensitive, especially when IMRT techniques are used.

6.4 Shielding Evaluation Report

The shielding evaluation report *should* contain the following information.

- Title page: Indicate the type of treatment unit, serial number, set of photon and electron beams available, location of facility, date of survey, and the name(s) of the person(s) who performed the survey and prepared the report.
- Methods: There *shall* be a description of: (1) the survey technique and instrumentation; (2) machine operating parameters for each set of measurements; and (3) methods used to obtain from the measurements the dose equivalent (H) for periods of 1 y, one week, and for uncontrolled locations, if applicable, "in-any-one-hour" of operation. Typically, tables of values are given of the workloads (W), for both the primary and leakage radiation, occupancy factors (T), and use factors (U) used for the evaluation. Separate workloads *should* be determined for each photon beam available. Special considerations for radiation shielding related to TBI and IMRT procedures *shall* be described. The table(s)

should include *W*, *U* and *T* values used and the type of area (controlled, uncontrolled or restricted) at selected locations outside each barrier.

- Floor plan and section views: Survey points are identified on plan and elevation drawings and they are correlated with the measurement results tables. All areas beyond the shielded vault *should* be identified with regard to type of use.
- Instruments: The type, model and serial number of each dosimeter used for the survey *shall* be indicated. The date of last calibration and the type and quality of radiation the instrument was calibrated for *shall* be stated.
- Results: Values of the maximum dose-equivalent rate outside all shielding barriers as a function of gantry angle *shall* be indicated along with the absorbed-dose rate at the isocenter during the measurements. The report *should* contain a table of the maximum dose equivalent to be expected annually or per week and in-any-one-hour of operation for all points surveyed.

Dose-equivalent measurements are accomplished using instruments that give either integrated or instantaneous readings at locations beyond primary or secondary barriers.

- *Integrated dose equivalent*. For a short-term, integrated dose-equivalent measurement at a location outside the protective barriers, the gantry orientation is fixed for the duration of the measurement and at primary-barrier locations, the measured value H_{MU} (the dose equivalent per monitor unit) results in a weekly TADR (R_w) given by Equation 6.1.

$$R_w = H_{MU}\, W(\text{MU})\, U \tag{6.1}$$

In Equation 6.1, *W*(MU) is the primary-beam workload in monitor units per week, rather than absorbed dose per week, and *U* is the use factor for the barrier shielding the location. The maximum dose equivalent in-any-one-hour is $R_h = (M/40)\, R_w$, assuming a 40 h workweek. Here *M* (Equation 3.14) is the ratio of the maximum number of procedures in an hour to the average number of procedures in an hour.

For secondary barriers, two short-term integrated measurements are needed, one with and one without a scattering phantom placed in the beam at isocenter. If the total short-term measured dose equivalent per monitor unit is $H_{MU,total}$ (made with the phantom in beam) and the result, without the phantom and the collimator closed, is $H_{MU,L}$ (leakage radiation only), then the phantom-scattered dose equivalent per monitor unit is $H_{MU,ps} = H_{MU,total} - H_{MU,L}$. The resulting weekly TADR is given in Equation 6.2.

$$R_w = [H_{MU,L} \; W_L(\text{MU})] + [H_{MU,\,ps} \; W_{ps}(\text{MU}) \; U_{ps}] \tag{6.2}$$

In Equation 6.2, W_L(MU) and W_{ps}(MU) are the leakage and patient-scattered-radiation workloads, respectively, in monitor units per week. The concepts of patient-scattered and leakage radiation workload are discussed in Sections 3.2.5.2 and 3.2.5.3.

The maximum dose equivalent in-any-one-hour (R_h) *should* also be assessed.

The dose equivalent to an individual at the location where R_w applies *shall* be multiplied by the occupancy factor (T) for that location to obtain a value for the dose equivalent ($H = R_w \; T$) to be compared to the shielding design goal (P).

- *Instantaneous dose-equivalent rate.* For measurements made with rate-mode instruments, the instantaneous dose-equivalent rate in sieverts per hour is denoted as *IDR*, while the accelerator production rate in monitor units per hour is denoted as $\dot{D}_0$(MU). For locations outside a primary barrier, the time averaged weekly dose-equivalent rate (R_w) is given in Equation 6.3 (see Equations 3.4 and 3.8).

$$R_w = \frac{IDR \; W(\text{MU}) \; U}{\dot{D}_0(\text{MU})} \tag{6.3}$$

The annual dose equivalent is 50 R_w, while the maximum dose equivalent in-any-one-hour is $R_h = (M/40) \; R_w$.

For total-radiation and leakage-radiation instantaneous dose-equivalent rates (IDR_{total} and IDR_L, respectively) measured at a location outside a secondary barrier, the resulting weekly TADR (R_w) is given in Equation 6.4.

$$R_w = \left[\frac{IDR_L \; W_L(\text{MU})}{\dot{D}_o(\text{MU})}\right] + \left[\frac{IDR_{ps} \; W_{ps}(\text{MU}) \; U_{ps}}{\dot{D}_o(\text{MU})}\right] \quad (6.4)$$

In Equation 6.4:

W_L = leakage-radiation workload (MU week^{-1}) (see Equation 3.6)

W_{ps} = patient-scattered-radiation workload (MU week^{-1}), which will normally be equal to the primary-barrier workload

U_{ps} = use factor for the gantry orientation used during the measurements

IDR_L = leakage-radiation instantaneous dose-equivalent rate

IDR_{ps} = patient-scattered radiation instantaneous dose-equivalent rate (as derived from IDR_{total} in Equation 3.10)

$\dot{D}_o$ (MU) = accelerator production rate (MU h^{-1})

- Conclusions and recommendations: If the shielding is acceptable, a statement indicating that the facility meets the requirements of the state or other regulatory body for radiation protection is made and the guidelines for radiation protection at medical facilities are quoted. If items of noncompliance are found, recommendations for correction of any problems are given and the need for a follow-up survey is indicated.

7. Examples

7.1 Conventional Treatment Unit with Maze

Figure 7.1[17] shows a proposed room layout for a dual-photon energy (6 MV low-energy and 18 MV high-energy) facility. The shielding design goals (P) are 0.02 mSv week^{-1} (1 mSv y^{-1}) for uncontrolled areas (*e.g.*, public access areas) and 0.1 mSv week^{-1} (5 mSv y^{-1}) for controlled areas (*e.g.*, in this case the treatment control area) (Section 1.4). The maximum dose equivalent in-any-one-hour is 0.02 mSv (20 μSv). The expected workload for 18 MV x rays is 30 patients per 8 h day, 5 d week^{-1} and an absorbed dose of 3 Gy delivered at the isocenter per patient. It is anticipated that an additional 15 patients per day will be treated with 6 MV x rays to the same absorbed dose. The accelerator has a maximum absorbed-dose output rate at 1 m of 12 Gy min^{-1}, and the normal rate used is 5 Gy min^{-1}. The isocenter is at 1 m from the radiation source for this facility. In each case, the examples will be presented first without IMRT considerations and then the changes for IMRT will follow, if applicable. The first example involving IMRT is in Section 7.1.4.

In the calculations, P for uncontrolled areas (0.02 mSv week^{-1}) is input as 20×10^{-6} Sv week^{-1}, and P for controlled areas (0.1 mSv week^{-1}) is input as 0.1×10^{-3} Sv week^{-1}, since the equations require the units in sievert. The results of the calculations for comparison with P and the limit for TADR in-any one-hour are given in μSv week^{-1} and microsievert in-any-one-hour, respectively.[18]

7.1.1 *Primary Barrier at Location C*

To determine the required barrier thickness at Location C, an unattended parking lot, Equation 2.1 is used, in which:

[17]Figure 7.1 is similar to Figure 2.7 in Section 2.4.1, which is for a general case. Figure 7.1 is more detailed and its notation is used throughout the Section 7 examples.

[18]See footnotes 5 and 6 in Section 2.2 concerning the quantities distance (d), workload (W), and transmission factor (B).

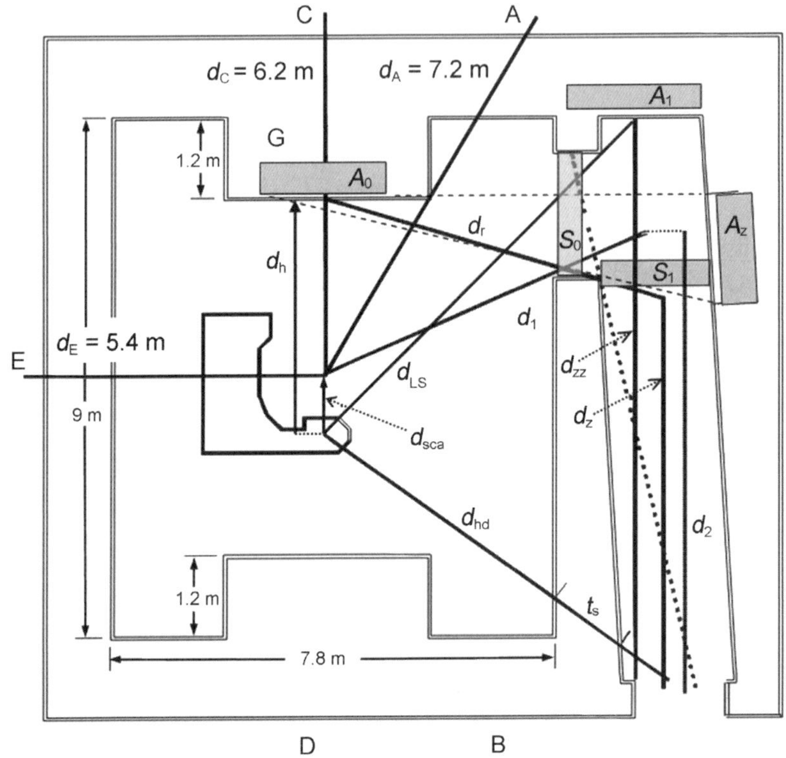

Fig. 7.1. Example for a dual-energy linear accelerator room with maze barrier.

P = shielding design goal for an uncontrolled area = 20×10^{-6} Sv week^{-1}

d_C = distance from the isocenter to 0.3 m beyond the barrier = 6.2 m

W(18 MV) = workload for a 5 d week = 450 Gy week^{-1} [(30 × 3 × 5) for the 18 MV primary-beam x rays] (Note: 6 MV patients are not included in this calculation since the higher energy and workload are from 18 MV. This assertion will be examined below.)

U = U_G = use factor for primary barrier G = 0.25

T = occupancy factor for the unattended parking lot = (1/40) = 0.025 (from Table B.1 in Appendix B)

$$B_{pri} = \frac{(20 \times 10^{-6})\,(6.2 + 1)^2}{(450)(0.25)(0.025)} = 3.69 \times 10^{-4}$$

The required number of *TVL*s to produce this attenuation is determined from Equation 2.2:

$$n = \log\left(\frac{1}{3.69 \times 10^{-4}}\right) = 3.43$$

From Equation 2.3 and the *TVL* values for 18 MV x rays in ordinary concrete from Table B.2 (TVL_1 = 45 cm, TVL_e = 43 cm)

$$t_{pri} = (45 \text{ cm}) + (3.43 - 1)(43 \text{ cm}) = 149.6 \text{ cm} \simeq 150 \text{ cm}$$

To determine whether this barrier thickness is adequate for the additional workload from 6 MV x rays, the following are used:

$$W(6 \text{ MV}) = (15 \text{ patients d}^{-1})(3 \text{ Gy patient}^{-1})(5 \text{ d week}^{-1})$$

$$= 225 \text{ Gy week}^{-1}$$

$$TVL_1 = 37 \text{ cm and } TVL_e = 33 \text{ cm}$$

The transmitted dose equivalent per week at Location C is obtained from Equation 2.1 with the shielding design goal (*P*) replaced by the dose equivalent (*H*) and with the primary-barrier transmission factor given by Equation 2.3.

$$H(6 \text{ MV}) = B_{pri} \; W \; U \; T \; (1 + d_C)^{-2}$$

$$= 10^{-\left\{1 + \left[\frac{(150 - 37)}{33}\right]\right\}} (225)(0.25)\left[\frac{0.025}{(7.2)^2}\right]$$

$$= 1 \times 10^{-6} \text{ Sv week}^{-1} = 1 \; \mu\text{Sv week}^{-1}$$

This 6 MV dose equivalent per week of 1 μSv is only 5 % of the shielding design goal and would not affect the primary-barrier thickness in this example. It must now be determined if the maximum dose equivalent in-any-one-hour limit is satisfied.

7.1.2 *Time Averaged Dose-Equivalent Rate Considerations at Location C*

The maximum dose equivalent in-any-one-hour is determined when the maximum absorbed-dose output rate at the isocenter (1 m from the source), which is $\dot{D}_o$ = 12 Gy min^{-1} or 720 Gy h^{-1}, is used. At this absorbed-dose output rate, the expected dose-equivalent rate per hour at Location C, with the transmission factor $B_{pri} = 3.69 \times 10^{-4}$, is:

$$IDR(18\ \text{MV}) = \frac{(720)(3.69 \times 10^{-4})}{(7.2)^2} = 5.1 \times 10^{-3}\ \text{Sv h}^{-1}$$

Applying Equation 3.8 for 18 MV x rays only, the weekly TADR at Location C is:

$$R_w(18\ \text{MV}) = \frac{IDR\ W_{pri}\ U_{pri}}{\dot{D}_0}$$

$$= (5.1 \times 10^{-3}\ \text{Sv h}^{-1})(450\ \text{Gy week}^{-1})(0.25)(720\ \text{Gy h}^{-1})^{-1}$$

$$= 8 \times 10^{-4}\ \text{Sv week}^{-1} = 800 \times 10^{-6}\ \text{Sv week}^{-1}$$

For 6 MV x rays the results are:

$$IDR(6\ \text{MV}) = (720)\ 10^{-\left\{1 + \left[\frac{(150-37)}{33}\right]\right\}}\ (7.2)^{-2}$$

$$= 5.2 \times 10^{-4}\ \text{Sv h}^{-1}$$

and

$$R_w(6\ \text{MV}) = (5.2 \times 10^{-4})(225)(0.25)(720)^{-1}$$

$$= 4.1 \times 10^{-5}\ \text{Sv week}^{-1} = 41 \times 10^{-6}\ \text{Sv week}^{-1}$$

A reasonable estimate is that no more than 10 patients can be treated in an hour. Since the average number of patients treated in an hour is 5.6 (45 patients d^{-1} / 8 h d^{-1}), the value of M needed for Equation 3.14 is 10/5.6 = 1.8.

Therefore, the dose equivalent in-any-one-hour from a combination of both 6 and 18 MV patients is:

$$R_h = \left(\frac{M}{40}\right) R_w(\text{total}) = \left(\frac{M}{40}\right) [R_w(6\ \text{MV}) + R_w(18\ \text{MV})]$$

$$= \left(\frac{M}{40}\right)[(41 \times 10^{-6}) + (800 \times 10^{-6})]$$

$$= \left(\frac{1.8}{40}\right)(841 \times 10^{-6}) = 38 \times 10^{-6}\ \text{Sv} = 38\ \mu\text{Sv in-any-one-hour}$$

This TADR value in-any-one-hour does not satisfy the 20 μSv in-any-one-hour requirement. At least 1 *HVL* (18 MV) of additional shielding is required for the primary barrier at Location C. Since 1 *HVL* just meets the TADR limit for the worst case of an all 18 MV workload, the conservatively safe choice is to add 2 *HVL*s.

Therefore, the new barrier thickness for the primary barrier protecting Location C is:

$$\begin{aligned} t_{pri}(\mathrm{C}) &= 150\ \mathrm{cm} + 2\ HVL(18\ \mathrm{MV}) \\ &= (150\ \mathrm{cm}) + (2)(0.301)\ TVL_e(18\ \mathrm{MV}) \\ &= (150\ \mathrm{cm}) + (2)(0.301)(43\ \mathrm{cm}) = 176\ \mathrm{cm} \end{aligned}$$

With the additional concrete, the final maximum dose equivalent in-any-one-hour is 9 μSv, well under the 20 μSv TADR limit.

A less conservatively safe alternative of imposing a restriction of not treating more than five patients in-any-one-hour would just satisfy the 20 μSv TADR limit but likely be difficult to maintain.

7.1.3 *Patient-Scattered Radiation Considerations at Location C*

Equation 2.7 is used to determine the barrier thickness required for shielding patient-scattered radiation in the vicinity of Location C. For this calculation, a position 10 degrees off the beam central axis is assumed. The worst case is when the beam is pointing toward Location C because the scatter fraction (a) is highest and the energy of the small angle scattered radiation is also the highest. If the primary barrier is sufficiently wide that it will also shield small angle scattered radiation, no additional thickness is needed to shield the scattered radiation. The following example illustrates the adequacy of the primary barrier for patient-scattered radiation.

The input data are:

$t_{pri}(\mathrm{C})$ = 176 cm
d_{sca} = 1 m
d_{sec} = 6.2 m, the distance from isocenter to Location C
a(18 MV) = 1.42×10^{-2} (Table B.4 for 18 MV scatter through 10 degrees at 2.5 cm depth)
a(6 MV) = 1.04×10^{-2} (Table B.4 for 6 MV scatter through 10 degrees at 2.5 cm depth)
W(18 MV) = 450 Gy week^{-1}

W(6 MV) = 225 Gy week^{-1}
T = 1/40 = 0.025
U = 0.25 for small angle scattered radiation with the gantry directed at Location C
F = (40 × 40) cm^2 (the maximum field size is used to be conservatively safe)
TVL_{sca}(6 MV at ≈ 10 degrees) = 35 cm of ordinary concrete (Table B.5a and interpolation)
TVL_{sca}(18 MV at ≈ 10 degrees) = 45 cm of ordinary concrete (Table B.5a)

The maximum transmitted patient-scattered dose equivalent at Location C from 6 MV x rays is:

$$B_{sca}(6\ \text{MV}) = 10^{-\left(\frac{176}{35}\right)} = 9.36 \times 10^{-6}$$

$$H_{sca}(6\ \text{MV})$$

$$= B_{sca}\, a(6\ \text{MV})(40 \times 40)(400)^{-1}\, W(6\ \text{MV})\ U\ T\ (d_{sec})^{-2}\ (d_{sca})^{-2}$$

$$= (9.36 \times 10^{-6})(1.04 \times 10^{-2})(40 \times 40)(400)^{-1}$$

$$(225)(0.25)(0.025)(6.2)^{-2}(1)^{-2}$$

$$= 14 \times 10^{-9}\ \text{Sv week}^{-1} = 0.014\ \mu\text{Sv week}^{-1}$$

and for 18 MV scattered radiation:

$$B_{sca}(18\ \text{MV}) = 10^{-\left(\frac{176}{45}\right)} = 1.23 \times 10^{-4}$$

$$H_{sca}(18\ \text{MV})$$

$$= B_{sca}\, a(18\ \text{MV})(40 \times 40)(400)^{-1} W(18\ \text{MV}) U\, T\, (d_{sec})^{-2}\ (d_{sca})^{-2}$$

$$= (1.23 \times 10^{-4})(1.42 \times 10^{-2})(40 \times 40)(400)^{-1}$$

$$(450)(0.25)(0.025)(6.2)^{-2}(1)^{-2}$$

$$= 5.1 \times 10^{-7}\ \text{Sv week}^{-1} = 0.51\ \mu\text{Sv week}^{-1}$$

Clearly, 6 MV patient-scattered radiation transmitted through the primary barrier is insignificant and the 18 MV patient-scattered-radiation contribution at Location C is well below the shielding design goal of 20 μSv week^{-1}. Therefore, the wall thickness determined for the primary barrier will be more than adequate to shield against scattered radiation. Note that for the average gantry angle, the use factor will be four times larger but the attenuation will be at least 100 times greater for the 90 degree (average angle) scattered radiation and the scatter fractions will be ~100 times smaller also.

7.1.4 *Leakage-Radiation Considerations at Location C*

To determine the barrier thickness required to shield against leakage radiation for Location C, Equations 2.3 and 2.8 are used. Because the head of a treatment unit is generally shielded to better than 0.1 % (the 10^{-3} factor), the primary barrier is adequate for shielding the additional radiation from leakage, as illustrated in the following calculations.

The input data values are:

P = 20×10^{-6} Sv week^{-1}

$t_{pri}(C)$ = 176 cm

d_L = 6.2 m

W(18 MV) = 450 Gy week^{-1}

W(6 MV) = 225 Gy week^{-1}

T = 1/40 = 0.025

TVL_L(18 MV) = 36 cm (TVL_1) and 34 cm (TVL_e) (approximate values from Table B.7)

TVL_L(6 MV) = 34 cm (TVL_1) and 29 cm (TVL_e) (approximate values from Table B.7)

The transmitted 6 MV leakage dose equivalent at Location C is:

$$H_L(6\text{ MV}) = B_L(6\text{ MV})\,(10^{-3})\,W(6\text{ MV})\,T\,(d_L)^{-2}$$

$$= 10^{-\left\{1+\left[\frac{(176-34)}{29}\right]\right\}}(10^{-3})(225)(0.025)(6.2)^{-2}$$

$$= 1.9 \times 10^{-4}\ \mu\text{Sv week}^{-1}$$

Note that the head-leakage radiation ratio (L_f) (Equation 2.12) for the low-energy x-ray beam in a dual-energy machine may be <0.1 % of the absorbed dose at 1 m from the useful beam. In the absence of specific data, the conservatively safe value $L_f = 10^{-3}$ is used.

For 18 MV:

$$H_L(18\text{ MV}) = B_L(18\text{ MV})\,(10^{-3})\,W(18\text{ MV})\,T\,(d_L)^{-2}$$

$$= 10^{-\left\{1+\left[\frac{(176-36)}{34}\right]\right\}}(10^{-3})(450)(0.025)(6.2)^{-2}$$

$$= 2.2 \times 10^{-3}\ \mu\text{Sv week}^{-1}$$

Neither of these dose equivalents is significant compared to the shielding design goal (20 μSv week^{-1}). Therefore, the leakage-shielding requirement is more than adequately met by the primary-barrier thickness.

IMRT modifications:

When allowing for IMRT procedures in Section 7.1, it is assumed that 80 % of the 6 MV x-ray patients and 40 % of the 18 MV patients will be treated using the IMRT technique. For either x-ray energy, experience on similar accelerators indicates that the average of C_I (for use in Equation 3.6) is about five. Furthermore, it is assumed that the prescribed total treatment absorbed dose per patient is unchanged by IMRT and that the number of patients treated with each energy is unchanged (30 and 15 d^{-1}, respectively). Using Equation 3.6 the leakage-radiation workload for 6 MV x rays is:

$$W_L(6\text{ MV}) = W_{\text{conv}} + C_I\,W_{\text{IMRT}}$$

$$= (0.2)(225\text{ Gy week}^{-1}) + (5)(0.8)(225\text{ Gy week}^{-1})$$

$$= (4.2)(225\text{ Gy week}^{-1}) = 945\text{ Gy week}^{-1}$$

Similarly, for 18 MV the leakage-radiation workload with IMRT is:

$$W_L(18\text{ MV}) = (0.6)(450) + (5)(0.4)(450)$$

$$= (2.6)(450) = 1,170\text{ Gy week}^{-1}$$

In this example for Location C, the transmitted leakage dose equivalent per week at Location C is increased by a factor of 4.2 (6 MV) and 2.6 (18 MV) relative to the conventional only procedures considered above. These increased rates are still insignificant (<0.1 μSv week^{-1}). This example demonstrates that the primary barrier is adequately shielded for leakage even with the increased leakage rates due to IMRT.

7.1.5 *Leakage- and Patient-Scattered-Radiation Considerations for Location A*

Here, only leakage and patient-scattered radiations are considered, since this is a secondary barrier and thus there is no primary radiation directed at Location A. Location A is 30 degrees off the beam centerline and, as a conservatively safe assumption, the minimum scatter angle of 30 degrees is used to look up the scatter fraction (a) from Table B.4. The input data to be used in Equations 2.2, 2.7 and 2.8 are:

P = 20×10^{-6} Sv week^{-1}, the shielding design goal for uncontrolled areas

T = 1/40 = 0.025

W(18 MV) = 450 Gy week^{-1}

W(6 MV) = 225 Gy week^{-1}

d_{sca} = 1 m

d_{sec} = 7.2 m, the distance from isocenter to Location A

a(18 MV) = 2.53×10^{-3} (for 18 MV at 30 degrees and 2.5 cm depth, Table B.4)

a(6 MV) = 2.77×10^{-3} (for 6 MV at 30 degrees and 2.5 cm depth, Table B.4)

F = (40×40) cm^2 (again, as a conservatively safe assumption)

TVL_{sca}(18 MV) = 32 cm of concrete (30 degree scatter, Table B.5a)

TVL_{sca}(6 MV) = 26 cm of concrete (30 degree scatter, Table B.5a)

d_L = 7.2 m, the distance from isocenter to Location A

TVL_L(18 MV leakage radiation) = 36 cm (TVL_1) and 34 cm (TVL_e) (Table B.7)

TVL_L(6 MV leakage radiation) = 34 cm (TVL_1) and 29 cm (TVL_e) (Table B.7)

For calculation of the barrier thickness necessary for *patient-scattered radiation*, Equation 2.7 is used with $U = 0.25$ and a(30 degrees). For 18 MV x rays:

$$B_{ps}(18\text{ MV}) = \frac{(20 \times 10^{-6})(1)(7.2)^2(400)}{(2.53 \times 10^{-3})(450)(0.25)(0.025)(40 \times 40)}$$

$$= 3.64 \times 10^{-2}$$

and from Equation 2.2:

$$n(\text{patient scatter, 18 MV}) = 1.44$$

The required barrier slant thickness ($t_{s,sca}$) for 18 MV scattered radiation is:

$$t_{s,sca}(18\text{ MV}) = (1.44)(32\text{ cm}) = 46.1\text{ cm} \simeq 46\text{ cm}$$

For 6 MV x rays:

$$B_{ps}(6\text{ MV}) = \frac{(20 \times 10^{-6})(1)(7.2)^2(400)}{(2.77 \times 10^{-3})(225)(0.25)(0.025)(40 \times 40)}$$

$$= 6.65 \times 10^{-2}$$

$$n(\text{patient scatter, 6 MV}) = 1.18$$

and

$$t_{s,sca}(6\text{ MV}) = (1.18)(26\text{ cm}) = 30.7\text{ cm} \simeq 31\text{ cm}$$

Combination of these two barrier requirements leads to adding 1 *HVL* (a conservatively safe value for 18 MV) to the larger of the two values. Thus:

$$t_{s,sca}(\text{total}) = (46) + (0.301)(32) = 55.6\text{ cm.}$$

For *leakage radiation* at 18 MV, using Equation 2.8:

$$B_L(18\text{ MV}) = \frac{(10^3)(20 \times 10^{-6})(7.2)^2}{(450)(0.025)} = 9.22 \times 10^{-2}$$

$$n(\text{leakage}) = 1.04$$

and

$$t_{s,L}(18\text{ MV}) = (1)(36\text{ cm}) + (0.04)(34\text{ cm}) = 37.3\text{ cm.}$$

For 6 MV:

$$B_L(6\text{ MV}) = \frac{(10^3)(20 \times 10^{-6})(7.2)^2}{(225)(0.025)} = 0.184$$

$$n(\text{leakage}) = 0.73$$

and

$$t_{s,L}(6\text{ MV}) = (0.73)(34\text{ cm}) = 25\text{ cm.}$$

Combining the requirements for leakage radiation from both energy components[19] gives:

$$t_{s,L}(\text{total}) = (37.2) + (0.301)(36) = 48\text{ cm.}$$

Note that in each case the first *TVL* was used rather than the equilibrium value since the calculated barrier slant thickness is only slightly thicker than a single *TVL*.

Finally, the scattered- and leakage-radiation barrier requirements (55.6 and 48 cm, respectively) are comparable and thus the total barrier slant thickness is given by the higher value plus one thickness of the highest *HVL*:

$$t_{s,Tot} = (55.6) + (0.301)(36) = 66.4\text{ cm.}$$

It would not be appropriate in this case to apply the obliquity factor of cos 30 degrees, since the patient-scattered and leakage

[19]As stated in Section 1.4.3, this use of the "two-source rule" for dual energies that cannot be used simultaneously represents a conservatively safe assumption.

radiations are not emanating from an apparent point source and hence cannot be assigned a unique angle of incidence to the barrier. Therefore, $t_{s,Tot}$ is taken as the barrier thickness (t_{Tot}).

In order to verify that the weekly TADR does not exceed the 20 μSv week^{-1} shielding design goal, the 66.4 cm barrier thickness is reapplied in the above equations to calculate the individual dose-equivalent components. The resulting values, in μSv week^{-1}, are ~2.8, 4.6 for the 18 MV and 0.8, 0.8 for the 6 MV, for the leakage- and patient-scattered-radiation components, respectively. Their sum is 9 μSv week^{-1}, which is well below the shielding design goal.

IMRT modifications:

For this example, IMRT procedures increase the leakage-radiation workloads for 18 MV and 6 MV to 1,170 Gy week^{-1} and 945 Gy week^{-1}, whereas the scattered-radiation workloads do not change with IMRT. The barrier slant thicknesses (t_s) are taken as the barrier thicknesses (t) (*e.g.*, $t_{s,L} = t_s$).

For 18 MV leakage radiation:

$$B_L(18\ \text{MV}) = 3.54 \times 10^{-2}$$

$$n(\text{leakage}) = 1.45$$

and the barrier thickness for this leakage-radiation component is:

$$t_L(18\ \text{MV}) = (36\ \text{cm}) + (0.45)(34\ \text{cm}) = 51.3\ \text{cm}.$$

For 6 MV x-ray leakage radiation:

$$t_L(6\ \text{MV}) = (34\ \text{cm}) + (0.36)(29\ \text{cm}) = 44.4\ \text{cm}.$$

The combined thickness is given by the higher value plus one added layer of the higher *HVL*:

$$t_L(\text{IMRT}) = (51.3) + (0.301)(36) = 62\ \text{cm}.$$

When combined with the comparable thickness required by the scattered radiation component (55.6 cm), the total wall thickness becomes:

$$t_{Tot} = (62) + (0.301)(36) = 73 \text{ cm.}$$

Verification that the total of dose-equivalent contributions per week at Location A is acceptable with a barrier thickness of 73 cm, yields these results after taking into account the occupancy factor of $T = 1/40$:

$$\text{Total from 18 MV scattered and leakage radiations} = 7.5\ \mu\text{Sv week}^{-1}$$

$$\text{Total from 6 MV scattered and leakage radiations} = 2.5\ \mu\text{Sv week}^{-1}$$

Thus, the combined total of 10 μSv week^{-1} is acceptable.

7.1.6 *Time Averaged Dose-Equivalent Rate (in-any-one-hour) Considerations for Location A*

Using the maximum absorbed-dose output rate at 1 m of 720 Gy h^{-1}, the expected dose-equivalent rate at Location A is the sum of the instantaneous dose-equivalent rates contributed by patient-scattered and leakage radiations (IDR_{ps} and IDR_L, respectively). Since the barrier path length is 66.4 cm, in the non-IMRT case, IDR_{ps} from 18 MV x rays at Location A is:

$$IDR_{ps}(18\text{ MV}) = \dot{D}_o\, a(18\text{ MV})\, F\, (400)^{-1}\, B_{sca}(18\text{ MV})\, (d_{sec})^{-2}$$

$$= \frac{(720)(2.53 \times 10^{-3})(40 \times 40)\ 10^{-\left(\frac{66.4}{32}\right)}}{(7.2)^2(400)} = 1.2 \times 10^{-3}\ \text{Sv h}^{-1}$$

and the IDR_L for 18 MV is:

$$IDR_L(18\text{ MV}) = \dot{D}_o\,(10^{-3})\, B_L(18\text{ MV})\,(d_L)^{-2}$$

$$= \frac{(720)\ 10^{-\left\{1 + \left[\frac{(66.4 - 36)}{34}\right]\right\}}}{(10^3)(7.2)^2} = 1.8 \times 10^{-4}\ \text{Sv h}^{-1}$$

The total (measurable) *IDR* (18 MV) at Location A from leakage and patient-scattered radiations is:

$$IDR(18\text{ MV}) = IDR_{ps}(18\text{ MV}) + IDR_L(18\text{ MV})$$
$$= (12 \times 10^{-4}) + (1.8 \times 10^{-4}) = 13.8 \times 10^{-4}\text{ Sv h}^{-1}$$

Similarly for 6 MV x rays:

$$IDR_{ps}(6\text{ MV})$$
$$= (720)(2.77 \times 10^{-3})(40 \times 40)\ 10^{-\left(\frac{66.4}{26.1}\right)}\ (7.2)^{-2}(400)^{-1}$$
$$= 4.3 \times 10^{-4}\text{ Sv h}^{-1}$$

and

$$IDR_L(6\text{ MV}) = (720)(10^{-3})\ 10^{-\left\{1+\left[\frac{(66.4-34)}{29}\right]\right\}}\ (7.2)^{-2}$$
$$= 1.1 \times 10^{-4}\text{ Sv h}^{-1}$$

for a total dose-equivalent rate at 6 MV of:

$$IDR(6\text{ MV}) = 5.4 \times 10^{-4}\text{ Sv h}^{-1}$$

Equation 3.9 gives the weekly TADR (R_w) for 18 MV x rays as:

$$R_w(18\text{ MV})$$
$$= \left[\frac{IDR_L(18\text{ MV})\ W_L(18\text{ MV})}{\dot{D}_o}\right] + \left[\frac{IDR_{ps}(18\text{ MV})\ W_{pri}(18\text{ MV})\ U}{\dot{D}_o}\right]$$
$$= [(1.8 \times 10^{-4})(450)(720)^{-1}] + [(12 \times 10^{-4})(450)(0.25)(720)^{-1}]$$
$$= 3 \times 10^{-4}\text{ Sv week}^{-1}$$

Similarly for 6 MV:

$$R_w(6\text{ MV}) = 0.7 \times 10^{-4}\text{ Sv week}^{-1}$$

The total TADR in-any-one-hour (Equation 3.14), using $M = 1.8$ from Section 7.1.2, is:

$$R_h = \left(\frac{M}{40}\right)[R_w(18\text{ MV}) + R_w(6\text{ MV})]$$

$$= \left(\frac{1.8}{40}\right)(3 \times 10^{-4} + 0.7 \times 10^{-4})$$

$$= 1.67 \times 10^{-5}\text{ Sv} = 16.7\ \mu\text{Sv in-any-one-hour}$$

Therefore, the maximum dose equivalent in-any-one-hour of 16.7 μSv is below the TADR limit of 20 μSv in-any-one-hour.

IMRT modifications:

The barrier slant thickness was increased to 73 cm due to IMRT procedures. Now, it is necessary to verify that the maximum dose equivalent in-any-one-hour at Location A is acceptable. For 18 MV with IMRT the values of IDR_L, IDR_{ps} and R_w become:

$$IDR_{ps}(18\text{ MV}) = 7.4 \times 10^{-4}\text{ Sv h}^{-1}$$

$$IDR_L(18\text{ MV}) = 1.1 \times 10^{-4}\text{ Sv h}^{-1}$$

and

$$R_w(18\text{ MV})$$

$$= \left[\frac{IDR_L(18\text{ MV})\ W_L(18\text{ MV})}{\dot{D}_o}\right] + \left[\frac{IDR_{ps}(18\text{ MV})\ W_{pri}(18\text{ MV})\ U}{\dot{D}_o}\right]$$

$$= [(1.1 \times 10^{-4})(1{,}170)(720)^{-1}] + [(7.4 \times 10^{-4})(450)(0.25)(720)^{-1}]$$

$$= 2.9 \times 10^{-4}\text{ Sv week}^{-1} = 290\ \mu\text{Sv week}^{-1}$$

For 6 MV:

$$IDR_{ps}(6\text{ MV}) = 2.4 \times 10^{-4}\text{ Sv h}^{-1}$$

$$IDR_L(6\text{ MV}) = 0.63 \times 10^{-4}\text{ Sv h}^{-1},$$

and

$$R_w(6\text{ MV})$$

$$= \left[\frac{IDR_L(6\text{ MV})\ W_L(6\text{ MV})}{\dot{D}_o}\right] + \left[\frac{IDR_{ps}(6\text{ MV})\ W_{pri}(6\text{ MV})\ U}{\dot{D}_o}\right]$$

$$= [(0.63 \times 10^{-4})(945)(720)^{-1}] + [(2.4 \times 10^{-4})(225)(0.25)(720)^{-1}]$$

$$= 1.01 \times 10^{-4}\text{ Sv week}^{-1} = 101\ \mu\text{Sv week}^{-1}$$

The above inputs to Equation 3.14 yield the TADR in-any-one-hour:

$$R_h = \left(\frac{M}{40}\right) R_w = \left(\frac{1.8}{40}\right)(290 + 101)$$

$$= 17.6\ \mu\text{Sv} < 20\ \mu\text{Sv in-any-one-hour}$$

This value of R_h demonstrates that the maximum dose equivalent in-any-one-hour at Location A is acceptable under IMRT treatment conditions.

7.1.7 *Primary Barrier at Location D in the Treatment Control Area*

For the treatment control area (a controlled area), the shielding design goal P is 5 mSv y^{-1}, or 0.1 mSv week^{-1} (0.1×10^{-3} Sv week^{-1}; 100 μSv week^{-1}). Other input data values for Equation 2.1 are:

d_D = 6.2 m the distance from Location D to the isocenter
W(18 MV) = 450 Gy week^{-1}
W(6 MV) = 225 Gy week^{-1}
U = 0.25
T = 1, radiation workers in a controlled area
TVL_1(18 MV) = 47 cm, TVL_e(18 MV) = 43 cm
TVL_1(6 MV) = 37 cm, TVL_e(6 MV) = 33 cm

Equations 2.1 and 2.2 yield:

$$B_{pri}(18\text{ MV}) = \frac{(0.1 \times 10^{-3})(6.2 + 1)^2}{(450)(0.25)(1)} = 4.61 \times 10^{-5}$$

and

$$n = \log\left(\frac{1}{4.61 \times 10^{-5}}\right) = 4.34$$

The required primary-barrier thickness is:

$$t(\mathrm{D}) = (45\ \mathrm{cm}) + (4.34 - 1)(43\ \mathrm{cm}) = 188.6\ \mathrm{cm} \simeq 189\ \mathrm{cm}$$

Whether this thickness is adequate for the 6 MV dose-equivalent contribution at Location D can be verified as follows:

$$H(6\ \mathrm{MV}) = B_{\mathrm{pri}}(6\ \mathrm{MV})\ W\ U\ T\ (d_{\mathrm{D}} + 1)^{-2}$$

$$= 10^{-\left\{1+\left[\frac{(189-37)}{33}\right]\right\}}(225)(0.25)(1)(6.2+1)^{-2}$$

$$= 2.7\ \mu\mathrm{Sv\ week}^{-1}$$

This is well below (<10 % of) the shielding design goal P = 100 μSv week^{-1}.

7.1.8 *Secondary Barrier at Location B*

This secondary-barrier location is similar to Location A (of Section 7.1.5) in terms of geometry. It is different in that Location B is a controlled area with full occupancy. The input data values are:

$P = 0.1 \times 10^{-3}$ Sv week^{-1}
$T = 1$
W(18 MV) = 450 Gy week^{-1}
W(6 MV) = 225 Gy week^{-1}
d_{sca} = 1 m
d_{sec} = 7.2 m
a(18 MV) = 2.53×10^{-3} (for 18 MV at 30 degrees and 2.5 cm depth, Table B.4)
a(6 MV) = 2.77×10^{-3} (for 6 MV at 30 degrees and 2.5 cm depth, TableB.4)
$F = (40 \times 40)$ cm^2
TVL_{sca}(18 MV) = 32 cm of concrete (30 degree scatter, Table B.5a, Appendix B)
TVL_{sca}(6 MV) = 26 cm of concrete (30 degree scatter, Table B.5a, Appendix B)
d_{L} = 7.2 m
TVL_{L}(18 MV) = 36 cm (TVL_1) and 34 cm (TVL_e) (Table B.7)
TVL_{L}(6 MV) = 34 cm (TVL_1) and 29 cm (TVL_e) (Table B.7)

As in Section 7.1.5, the calculated barrier slant thicknesses are taken as the barrier thicknesses (*e.g.*, $t_{s,sca} = t_{sca}$). For calculation of the barrier thickness against *patient-scattered radiation,* Equation 2.7 is used with $U = 0.25$ and a(30 degrees). For 18 MV x rays:

$$B_{ps}(18\text{ MV}) = \frac{(0.1 \times 10^{-3})(1)(7.2)^2(400)}{(2.53 \times 10^{-3})(450)(0.25)(1)(40 \times 40)} = 4.55 \times 10^{-3}$$

$$n(\text{patient scatter, 18 MV}) = 2.34$$

The required barrier thickness for 18 MV scattered radiation is:

$$t_{sca}(18\text{ MV}) = (2.34)(32\text{ cm}) = 74.9\text{ cm}$$

For 6 MV x rays:

$$B_{ps}(6\text{ MV}) = \frac{(0.1 \times 10^{-3})(1)(7.2)^2(400)}{(2.77 \times 10^{-3})(225)(0.25)(1)(40 \times 40)} = 8.32 \times 10^{-3}$$

$$n(\text{patient scatter, 6 MV}) = 2.08$$

and

$$t_{sca}(6\text{ MV}) = (2.08)(26\text{ cm}) = 54.1\text{ cm}$$

For *leakage radiation* at 18 MV, using Equation 2.8:

$$B_L(18\text{ MV}) = \frac{(10^3)(0.1 \times 10^{-3})\,(7.2)^2}{(450)(1)} = 1.15 \times 10^{-2}$$

$$n(\text{leakage}) = 1.94$$

and

$$t_L(18\text{ MV}) = (36\text{ cm}) + (0.94)(34\text{ cm}) = 68\text{ cm}$$

For 6 MV leakage radiation:

$$B_L(6\text{ MV}) = \frac{(10^3)(0.1 \times 10^{-3})\,(7.2)^2}{(225)(1)} = 2.30 \times 10^{-2}$$

$$n(\text{leakage}) = 1.64$$

and

$$t_L(6\ \text{MV}) = (34\ \text{cm}) + (0.64)(29\ \text{cm}) = 52.6\ \text{cm}$$

For Location B, the combination of the patient-scattered-radiation thicknesses, by adding 1 *HVL* to the higher value since they are within 1 *TVL* of each other, gives:

$$t_{sca} = (74.9) + (0.301)(32) = 84.5\ \text{cm},$$

and combining the leakage-radiation values similarly gives:

$$t_L = (68) + (0.301)(34) = 78.2\ \text{cm}.$$

And the final barrier thickness then results from an additional application of the same two-source rule for the patient-scattered and leakage-radiation sources:

$$t_{Tot} = (84.5) + (0.301)(34) = 94.7\ \text{cm}$$

The adequacy of the calculated barrier thickness for all contributions combined should be verified. Using the thickness of 94.7 cm, the weekly dose-equivalent contributions from 18 MV scattered and leakage radiations at Location B are 24.1 and 16.3 μSv week^{-1}, respectively. For 6 MV, the values are 2.7 μSv week^{-1}, from scattered radiation, and 3.5 μSv week^{-1} from leakage radiation. The sum of these is 47 μSv week^{-1}, which is less than the shielding design goal (100 μSv week^{-1}).

IMRT modifications:

For this example, IMRT procedures increase the leakage-radiation workloads for 18 and 6 MV to 1,170 and 945 Gy week^{-1}, respectively, whereas the patient-scattered-radiation workloads do not change with IMRT.

In this case, for 18 MV leakage radiation:

$$B_L(18\ \text{MV}) = P\, d_L^2\, (10^3)\, (W_L\, T)^{-1} = 4.43 \times 10^{-3}$$

$$n(\text{leakage}) = 2.35,$$

and the barrier thickness:

$$t_L(18\ \text{MV}) = (36\ \text{cm}) + (1.35)(34)\ \text{cm} = 82\ \text{cm}.$$

In a similar manner, for 6 MV x-ray leakage radiation:

$$B_L(6\ \text{MV}) = 5.49 \times 10^{-3}$$

$$n(\text{leakage}) = 2.26$$

and

$$t_L(6\ \text{MV}) = (34\ \text{cm}) + (1.26)(29)\ \text{cm} = 70.5\ \text{cm}.$$

Since the patient-scattered radiation does not change with IMRT, the required barriers for these components remain 74.9 cm for 18 MV and 54.1 cm for 6 MV. Application of the two-source rule for the leakage- and scattered-radiation components gives 92.2 cm [(82) + (0.301) (34)] and 84.5 cm [(74.9) + (0.301) (32)], respectively. When these are combined the result is 102.4 cm [(92.2) + (0.301) (34)] for the secondary barrier at Location B when IMRT is used.

Verification that the total of dose-equivalent contributions per week at Location B are acceptable with a barrier thickness of 102.4 cm, yields these results:

$$\text{Total from 18 MV scattered and leakage radiations} = 39.1\ \mu\text{Sv week}^{-1}$$

$$\text{Total from 6 MV scattered and leakage radiations} = 9.4\ \mu\text{Sv week}^{-1}$$

Thus, the combined total of 48.5 μSv week^{-1} is acceptable.

7.1.9 *Secondary Barrier at Location E*

If, as in this example, the barriers are to be constructed from concrete, then it is generally preferable to make them a uniform thickness. Thus, only the barrier thickness at Location E needs to be considered for this wall, since Location E is at the minimum distance and the obliquity angle is zero (*i.e.*, $t_s = t$).

Location E is in a film reading area for the radiation oncologists and it runs along the entire barrier. Therefore, the qualified expert considers this area uncontrolled and has assigned an occupancy factor T = 1/5 (since individual radiation oncologists rarely spend more than 1 h d^{-1} at the film viewer). The shielding design goal is P = 20 μSv week^{-1}. Though the radiation oncologists working in this area may be considered radiation workers and assigned radiation dosimeters, the area is a hallway, hence the decision to call this an uncontrolled area.

The other input data values used in Equations 2.2, 2.7 and 2.8 are:

P = 20 × 10^{-6} Sv week^{-1}

d_{sca} = 1 m

d_{sec} = 5.4 m, distance from the patient (*i.e.*, the scattering object) at isocenter to Location E (30 cm beyond barrier)

a(18 MV) = 1.89 × 10^{-4}, for 90 degree scatter from Table B.4

a(6 MV) = 4.26 × 10^{-4}, for 90 degree scatter from Table B.4

W(18 MV) = 450 Gy week^{-1}

W(6 MV) = 225 Gy week^{-1}

F = (40 × 40) cm^2

d_L = 5.4 m, distance from isocenter to Location E for leakage-radiation calculations

TVL_L(18 MV leakage radiation) = 36 cm (TVL_1) and 34 cm (TVL_e) (Table B.7)

TVL_L(6 MV leakage radiation) = 34 cm (TVL_1) and 29 cm (TVL_e) (Table B.7)

TVL_{sca}(18 MV) = 19 cm of concrete (90 degree scatter, Table B.5a)

TVL_{sca}(6 MV) = 17 cm of concrete (90 degree scatter, Table B.5a)

For calculation of the barrier thickness for *patient-scattered radiation,* Equation 2.7 is used with U = 1 and a(90 degrees). For 18 MV x rays:

$$B_{ps}(18\text{ MV}) = \frac{(20 \times 10^{-6})(1)(5.4)^2(400)}{(1.89 \times 10^{-4})(450)(0.20)(40 \times 40)} = 8.57 \times 10^{-3}$$

$$n(\text{patient scatter, 18 MV}) = 2.07$$

The required barrier thickness for 18 MV scattered radiation is:

$$t_{sca}(18\text{ MV}) = (2.07)(19\text{ cm}) = 39.3\text{ cm}$$

For 6 MV x rays:

$$B_{ps}(6\text{ MV}) = \frac{(20 \times 10^{-6})(1)(5.4)^2(400)}{(4.26 \times 10^{-4})(225)(0.20)(40 \times 40)} = 7.6 \times 10^{-3}$$

$$n(\text{patient scatter, 6 MV}) = 2.12$$

and

$$t_{sca}(6\text{ MV}) = (2.12)(17\text{ cm}) = 36\text{ cm}$$

For *leakage radiation* at 18 MV, using Equation 2.8:

$$B_L(18\text{ MV}) = \frac{(10^3)(20 \times 10^{-6})(5.4)^2}{(450)(0.20)} = 6.48 \times 10^{-3}$$

$$n(\text{leakage}) = 2.19$$

and

$$t_L(18\text{ MV}) = (36\text{ cm}) + (1.19)(34\text{ cm}) = 76.5\text{ cm}$$

And for 6 MV:

$$B_L(6\text{ MV}) = \frac{(10^3)(20 \times 10^{-6})(5.4)^2}{(225)(0.20)} = 1.3 \times 10^{-2}$$

$$n(\text{leakage}) = 1.89$$

and

$$t_L(6\text{ MV}) = (34\text{ cm}) + (0.89)(29\text{ cm}) = 59.8\text{ cm}$$

Examining the computed barrier thicknesses in this example, it is observed that the leakage-radiation thicknesses are thicker than scattered radiation thicknesses (due mainly to the reduced scattered-radiation *TVL*s at 90 degrees). The combined scattered-radiation barrier requirement is ~45 cm, while the combined leakage-radiation requirement is ~87 cm. Since these are more than a *TVL* (36 cm) different, the higher value (87 cm) is used for the final barrier thickness. By calculation, it can be verified that a barrier thickness of 87 cm achieves the shielding design goal considering all contributions. The results are 8.8 μSv week^{-1} coming from the 18 MV mode leakage radiation, and 2.4 μSv week^{-1} from the 6 MV mode, taking into account the occupancy factor of T = 1/5. The total of 11.2 μSv week^{-1} is less than the shielding design goal of 20 μSv week^{-1}.

IMRT modifications:

With the increased leakage-radiation workload for 18 MV x rays of 1,170 Gy week^{-1}, the barrier transmission factor is:

$$B_L(18\ \text{MV}) = 2.49 \times 10^{-3}$$

$$n(\text{leakage}) = 2.60$$

and the barrier thickness:

$$t_L(18\ \text{MV}) = (36\ \text{cm}) + (1.60)(34\ \text{cm}) = 90.4\ \text{cm}$$

For 6 MV x-ray leakage radiation (W_L = 945 Gy week^{-1}):

$$t_L(6\ \text{MV}) = (34\ \text{cm}) + (1.51)(29\ \text{cm}) = 78\ \text{cm}$$

Since the scattered-radiation contributions are negligible and the barrier thickness computed for 18 MV leakage radiation (90.4 cm) is <1 *TVL* higher than the 6 MV thickness (78 cm), 1 *HVL* (10.8 cm) is added to the former. The result is a barrier of 101.2 cm thickness, which is 14.2 cm thicker than found without IMRT.

The combined dose-equivalent contributions from leakage radiation per week at Location E are computed to be 12.8 μSv week^{-1}, less than the 20 μSv week^{-1} shielding design goal.

It is necessary to determine if the TADR limit for the maximum dose equivalent in-any-one-hour (20 μSv) is achieved with this barrier thickness (101.2 cm). This assessment is made with R_h, the TADR in-any-one-hour. In this case, the scattered-radiation contributions are insignificant at Location E, so only the leakage-radiation contributions are computed. The results including IMRT contributions for 6 and 18 MV are:

$$R_w(6\text{ MV}) = IDR_L\ W_L(6\text{ MV})\ (\dot{D}_o)^{-1}$$

$$= [\dot{D}_o\ (10^{-3})\ B_L(6\text{ MV})\ (d_L)^{-2}]\ [W_L(6\text{ MV})\ (\dot{D}_o)^{-1}]$$

$$= (10^{-3})\ 10^{-\left\{1+\left[\frac{(101.2-34)}{29}\right]\right\}}\ (5.4)^{-2}\ (945)$$

$$= 15.6 \times 10^{-6}\text{ Sv week}^{-1}$$

$$R_w(18\text{ MV}) = 48.5 \times 10^{-6}\text{ Sv week}^{-1}$$

and, therefore

$$R_h = \left(\frac{M}{40}\right)[R_w(6\text{ MV}) + R_w(18\text{ MV})]$$

$$= \left(\frac{1.8}{40}\right)[(15.6 + 48.5) \times 10^{-6}]$$

$$= 3 \times 10^{-6}\text{ Sv} = 3\ \mu\text{Sv in-any-one-hour}$$

These results for the maximum dose equivalent in-any-one-hour are well below the TADR limit of 20 μSv.

7.1.10 *Leakage and Scattered Radiation at the Maze Door*

For a high-energy accelerator, the contributions of leakage and scattered radiations reaching the maze door are generally relatively low compared with the neutron capture gamma-ray and neutron dose-equivalent components examined in Sections 7.1.11 and 7.1.12. However, this is not always the case, and an example will be examined here.

The room layout is shown in Figure 7.1 and Equations 2.13 and 2.14 may be used to simplify the calculation for the dose equivalent at the door H_d.

$$H_d = f\,H_S + H_{LS} + H_{ps} + H_{LT}$$

where the components are calculated as given in Sections 7.1.10.1 through 7.1.10.6.

7.1.10.1 *Wall-Scattered Radiation Component, H_S.* The input data values used in the Equation 2.9 are:

W(18 MV) = 450 Gy week^{-1} and W(6 MV) = 225 Gy week^{-1}
U_G = 0.25

The distances as shown in Figure 7.1 are:

d_G = 4.2 m, corresponds to the distance d_{pp} + 1 m in Figure 2.7 and d_h in Equation 2.9
d_r = 5.9 m
d_z = 6.8 m

The wall reflection coefficients are:

α_G(18 MV) = 1.6 × 10^{-3} (Table B.8a, normal incidence, 75 degree angle of reflection)
α_G(6 MV) = 2.7 × 10^{-3} (Table B.8a, normal incidence, 75 degree angle of reflection)
α_z = 8 × 10^{-3} (Table B.8a, normal incidence, 75 degree angle of reflection, 0.5 MeV)
A_0 = 2.82 m^2, the maximum field size (40 × 40) cm^2 projected onto Wall G corresponding to Area A_0 in Figure 2.7 [*i.e.*, (168 × 168) cm^2]
A_z = 8.4 m^2, the cross-sectional area of the inner maze entrance as projected from the irradiated primary-beam area (A_0)

Equation 2.9 gives the value of H_S.

$$H_S = \frac{[(450)(1.6 \times 10^{-3}) + (225)(2.7 \times 10^{-3})]\ (0.25)(2.82)(8 \times 10^{-3})(8.4)}{[(4.2)(5.9)(6.8)]^2}$$

$$= 2.21 \times 10^{-6}\ \text{Sv week}^{-1} = 2.2\ \mu\text{Sv week}^{-1}$$

7.1.10.2 *Head-Leakage Wall-Scattered Radiation Component, H_{LS}.* Input data values used in Equation 2.10 are as follows:

L_f = 1 × 10^{-3} (assumed the same for both 6 and 18 MV x rays)
W_L(18 MV) = 450 Gy week^{-1} (in absence of IMRT enhancements)
W_L(6 MV) = 225 Gy week^{-1}

U_G = 0.25
α_1 (18 MV) = 4.5 × 10^{-3} (Table B.8b, 45 degree incidence, zero degree reflection angle, 18 MV)
α_1 (6 MV) = 6.4 × 10^{-3} (Table B.8b, 45 degree incidence, zero degree reflection angle, 6 MV)
A_1 = 2.8 m × 4.2 m = 11.8 m^2, the area of the wall
d_{LS} = 7.9 m, the distance from x-ray target to the maze centerline in Area A_1
d_{zz} = 9.9 m

The combined mode dose equivalent at the maze door from head-leakage radiation scattered by the Wall A_1 is:

$$H_{LS} = \frac{(10^{-3}) <W\alpha> (0.25)(11.8)}{[(7.9)(9.9)]^2} = 1.7 \times 10^{-6} \text{ Sv week}^{-1}$$

$$= 1.7 \ \mu\text{Sv week}^{-1}$$

Here:

$$<W\alpha> = [W_L(6 \text{ MV}) \ \alpha_1(6 \text{ MV})] + [W_L(18 \text{ MV}) \ \alpha_1(18 \text{ MV})]$$

$$= 3.47 \text{ Gy week}^{-1}$$

7.1.10.3 *Patient-Scattered Radiation Component, H_{ps}.* Equation 2.11 is used to determine the dose equivalent at the door, scattered by the patient with the beam pointing at Wall G, or Location C, as shown in Figure 7.1. The input data values are:

W(18 MV) = 450 Gy week^{-1}
W(6 MV) = 225 Gy week^{-1}
U_G = 0.25
F = (40 × 40) cm^2
d_{sca} = 1 m
d_{sec} = 7.3 m, the distance from isocenter to Area A_1
d_{zz} = 9.9 m, the distance from Area A_1 to door
A_1 = 11.8 m^2
a(6 MV) = 1.39 × 10^{-3}, the 6 MV scatter fraction at 45 degree scatter angle (Table B.4)
a(18 MV) = 8.64 × 10^{-4}, the 18 MV scatter fraction at 45 degree scatter angle (Table B.4)

α_1 = 2.2×10^{-2}, the concrete wall reflection coefficient for incident angle 45 degrees and reflection angle zero degree for 0.5 MeV monoenergetic photons (Table B.8b)

The concrete wall reflection coefficient α_1 is a function of the incident beam energy and the incident angle. After being scattered by the patient, the radiation energy can be as low as 0.5 MeV due to Compton interactions. Table B.8a demonstrates that the reflection coefficient decreases as the energy increases so using the 0.5 MeV coefficient will not underestimate the dose equivalent to the maze door.

Using the input data values shown above, Equation 2.11 is evaluated for both the 6 and 18 MV scattered radiations combined to obtain H_{ps}:

$$H_{ps} = \frac{<Wa> (0.25)\left(\frac{40 \times 40}{400}\right)(2.2 \times 10^{-2})(11.8)}{[(1)(7.3)(9.9)]^2}$$

$$= 34.9 \times 10^{-6} \text{ Sv week}^{-1} = 34.9 \ \mu\text{Sv week}^{-1}$$

Here, for simplicity:

$$<Wa> = [W_L(6 \text{ MV})\ a(6 \text{ MV})] + [W_L(18 \text{ MV})\ a(18 \text{ MV})]$$

$$= 0.702 \text{ Gy week}^{-1}$$

7.1.10.4 *Head-Leakage Radiation Through Maze Wall, H_{LT}.* The input data values used in Equation 2.12 are:

L_f = 1×10^{-3}

$W_L(18 \text{ MV})$ = 450 Gy week^{-1} and $W_L(6 \text{ MV})$ = 225 Gy week^{-1}

U_G = 0.25 (use factor for horizontal gantry orientation)

d_L = d_{hd} = 7.1 m (head-to-door distance for gantry orientation in Figure 7.1)

The oblique maze wall slant thickness (t_s) is 125 cm (Section 7.1.17) of concrete for the gantry orientation indicated in Figure 7.1. Since TVL_1 and TVL_e for 18 MV leakage radiation in concrete are 36 and 34 cm, respectively, the barrier transmission B (18 MV) is:

$$B(18\ \text{MV}) = 10^{-\left\{1+\left[\frac{(125-36)}{34}\right]\right\}} = 2.41 \times 10^{-4}$$

The dose equivalent H_{LT} from Equation 2.5 is:

$$H_{LT}(18\ \text{MV}) = (2.41 \times 10^{-4})(450)(0.25)(10^{-3})(7.3)^{-2}$$

$$= 5.14 \times 10^{-7}\ \text{Sv week}^{-1} = 0.5\ \mu\text{Sv week}^{-1}$$

Similarly, for transmitted 6 MV leakage radiation, $H_{LT}(6\ \text{MV})$ = 0.08 μSv week^{-1}. Therefore, the approximate total transmitted leakage radiation is H_{LT} = 0.6 μSv week^{-1}.

7.1.10.5 *Total Dose Equivalent Due to Scattered and Leakage Radiations,* H_{Tot}. The total photon dose equivalent at the maze door due to scattered and leakage radiations is, using Equation 2.14, and calculating f = 0.34 from the relative workloads and depth doses (absorbed dose) of the 6 and 18 MV beams:

$$H_{Tot} = 2.64\ H_G = 2.64\ (f\,H_S + H_{LS} + H_{ps} + H_{LT})$$

$$= 2.64\ [(0.34)(2.1) + 1.7 + 34.9 + 0.6] \times 10^{-6}$$

$$= 1 \times 10^{-4}\ \text{Sv week}^{-1} = 100\ \mu\text{Sv week}^{-1}$$

7.1.10.6 *IMRT Modifications*. There are changes in the leakage-radiation transmission through the maze barrier and the leakage- and scattered-radiation components arriving at the door. Patient-scattered radiation and primary-beam wall-scattered radiation are not affected by IMRT.

7.1.10.6.1 *Head-leakage wall-scattered-radiation component,* H_{LS}. Since only the leakage-radiation workload changes, the input data for Equation 2.10 are: W_L = 945 and 1,170 Gy week^{-1} for 6 and 18 MV, respectively. The dose equivalent at the maze door from head-leakage radiation scattered by Area A_1 becomes:

$$H_{LS} = 5.5 \times 10^{-6}\ \text{Sv week}^{-1}$$

7.1.10.6.2 *Head-leakage radiation transmitted through maze wall,* H_{LT}. For 18 MV, $H_{LT}(18\ \text{MV}) = 1.4 \times 10^{-6}$ Sv week^{-1} and for 6 MV, $H_{LT}(6\ \text{MV}) = 0.3 \times 10^{-6}$ Sv week^{-1}, yielding a combined total H_{LT} = 1.7 μSv week^{-1}.

The total dose equivalent at the door due to scattered and leakage radiations, H_{Tot} is:

$$H_{Tot} = 2.64\ H_G = 2.64\ (f\,H_S + H_{LS} + H_{ps} + H_{LT})$$
$$= 2.64\ [(0.34)(2.1) + 5.5 + 34.9 + 1.7] \times 10^{-6}$$
$$= 113 \times 10^{-6}\ \text{Sv week}^{-1} = 113\ \mu\text{Sv week}^{-1}$$

7.1.11 *Neutron Capture Gamma-Ray Dose Equivalent at the Maze Door*

Photoneutron production and, hence, the neutron capture gamma-ray dose equivalent is proportional to the leakage-radiation workload of high-energy x rays [W_L(18 MV) in this example]. To determine the neutron capture gamma-ray dose equivalent (h_φ), Equation 2.15 is used.

The total neutron fluence φ_A at the inner maze point is first determined using Equation 2.16. If the accelerator is a 18 MV machine (Varian Model 1800), then from Table B.9, $Q_n = 1.22 \times 10^{12}$ neutrons per x-ray gray at isocenter. The accelerator head is lead, therefore $a = 1$. The distance from the isocenter to the inner maze point (d_1) is 6.4 m. The length of the maze from the inner maze point to the door (d_2) is 8.5 m. The concrete room dimensions are shown in Figures 7.1 and 7.2.

The room surface area S_r for use in Equation 2.16 is the sum of the areas of ceiling and floor, front and back walls, and left and right walls. The average room height is 3.65 m, the average width is 7.8 m, and the average length is 7.8 m. The surface area of the room is therefore:

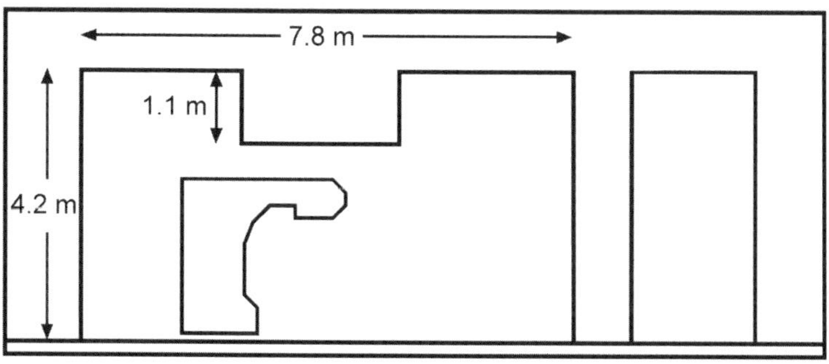

Fig. 7.2. Sectional diagram (through isocenter) of the treatment room.

$$S_r = 2[(7.8)(3.65) + (7.8)(3.65) + (7.8)(7.8)]$$

$$= 236 \text{ m}^2$$

The total neutron fluence per isocenter x-ray gray at the inner maze point (located where d_1 and d_2 intersect in Figure 2.8) from Equation 2.16 is:

$$\varphi_A = \frac{(1.22 \times 10^{12})}{(4\pi)(6.4)^2} + \frac{(5.4)(1.22 \times 10^{12})}{(2\pi)(236)} + \frac{(1.3)(1.22 \times 10^{12})}{(2\pi)(236)}$$

$$= 7.88 \times 10^9 \text{ neutron m}^{-2}$$

The neutron capture gamma-ray dose equivalent at the maze door, from Equation 2.15, is:

$$h_\varphi = K \, \varphi_A \, 10^{-\left(\frac{d_2}{TVD_2}\right)}$$

$$h_\varphi = (6.9 \times 10^{-16})(7.88 \times 10^9)10^{-\left(\frac{8.5}{5.4}\right)}$$

$$= 1.45 \times 10^{-7} \text{ Sv Gy}^{-1} (\textit{i.e.}, \text{ per gray at isocenter})$$

The weekly neutron capture gamma-ray dose equivalent at the maze door H_{cg} is, from Equation 2.17:

$$H_{cg} = W_L(18 \text{ MV}) \, h_\varphi = (450)(1.45 \times 10^{-7})$$

$$= 6.53 \times 10^{-5} \text{ Sv week}^{-1}$$

$$= 65.3 \text{ } \mu\text{Sv week}^{-1}$$

This is comparable to the total dose equivalent from x-ray scattered and leakage radiations.

IMRT modifications:

The 18 MV leakage-radiation workload with IMRT is 1,170 Gy week^{-1}, which is 2.6 times higher than without IMRT and, therefore, the neutron production is 2.6 times higher with IMRT than without. The weekly neutron capture gamma-ray dose equivalent at the maze door (H_{cg}), from Equation 2.17 becomes:

$$H_{cg} = 170 \text{ } \mu\text{Sv week}^{-1}$$

7.1.12 *Neutron Dose Equivalent at the Maze Door*

Two methods will be used to estimate the neutron dose equivalent at the maze entrance. For the treatment room described in Section 7.1.10, the areas of the inner maze opening S_0, and the cross-sectional area of the maze S_1, as shown in Figure 7.1, are needed for both methods. The values for this room are $S_0 = 9.2\ \text{m}^2$ and $S_1 = 8.4\ \text{m}^2$.

The neutron dose equivalent at the outer maze entrance ($H_{n,D}$) will first be determined using the method of Kersey (1979), Equation 2.18, input data values listed in Section 7.1.10, and the highest value of H_0 for an 18 MV x-ray treatment unit from Table B.9:

$$H_{\text{n,D}} = H_0 \left(\frac{S_0}{S_1}\right)\left(\frac{d_0}{d_1}\right)^2 10^{-\left(\frac{d_2}{5}\right)}$$

$$= (1.6 \times 10^{-3})\left(\frac{9.2}{8.4}\right)\left(\frac{1.41}{6.4}\right)^2 10^{-\left(\frac{8.5}{5}\right)}$$

$$= 1.7 \times 10^{-6}\ \text{Sv Gy}^{-1}\ (\textit{i.e.}, \text{per gray at isocenter})$$

Alternatively, to estimate the neutron dose equivalent using the method by Wu and McGinley (2003), the *TVD* for the maze is first determined using Equation 2.20:

$$TVD = 2.06\sqrt{(S_1)}$$

$$= 2.06\sqrt{(8.4)} = 6\ \text{m}$$

The neutron dose equivalent at the maze entrance is then determined using Equation 2.19, input data values listed in Section 7.1.11, and the value of φ_A obtained in that example:

$$H_{\text{n},D} = (2.4 \times 10^{-15})\ \varphi_A \sqrt{\frac{S_0}{S_1}} \left[1.64 \times 10^{-\left(\frac{d_2}{1.9}\right)} + 10^{-\left(\frac{d_2}{TVD}\right)}\right]$$

$$= (2.4 \times 10^{-15})(7.88 \times 10^9)\sqrt{\frac{9.2}{8.4}} \left[1.64 \times 10^{-\left(\frac{8.5}{1.9}\right)} + 10^{-\left(\frac{8.5}{6}\right)}\right]$$

$$= 0.8 \times 10^{-6}\ \text{Sv Gy}^{-1}\ (\textit{i.e.}, \text{per gray at isocenter})$$

This value is 46 % of what Kersey's method predicts. For the room design as shown, the alternative method is expected to give a more accurate estimate than the Kersey method. A conservatively safe approach would use the larger value for the neutron dose equivalent at the maze door:

$$H_n = W_L(18\text{ MV})\, H_{n,D} = (450\text{ Gy week})^{-1}(1.7 \times 10^{-6}\text{ Sv Gy}^{-1})$$

$$= 7.65 \times 10^{-4}\text{ Sv week}^{-1} = 765\ \mu\text{Sv week}^{-1}$$

In summary, using a conservatively safe estimate, the total dose equivalent (from neutrons and photons) at the maze door per week is:

$$H_w = H_{Tot} + H_{cg} + H_n = (100 + 65.3 + 765)\ \mu\text{Sv week}^{-1}$$

$$= 930\ \mu\text{Sv week}^{-1} \simeq 1\text{ mSv week}^{-1}$$

IMRT modifications:

Taking the conservatively safe approach, using Kersey's method for neutron dose equivalent per unit of x-ray absorbed dose, the total neutron dose equivalent per week at the door with IMRT is:

$$H_n = W_L\, H_{n,D} = (1{,}170)(1.7 \times 10^{-6}) = 1{,}989\ \mu\text{Sv week}^{-1}$$

The IMRT values of the photon contributions, H_{Tot} and H_{cg}, are 113 and 170 μSv week^{-1}, respectively. Thus, the combined photon and neutron dose equivalent at the maze door location with IMRT is H_w = 2,272 μSv week^{-1}.

7.1.13 *Shielding Barrier for the Maze Door*

The maze entrance is located in a controlled area and the shielding design goal is P = 0.1 mSv week^{-1}. For this example, the total dose equivalent at the door is 0.93 mSv week^{-1}, of which ~83 % is from neutrons, 10 % is from low-energy scattered and transmitted leakage photons, and 7 % is from neutron capture gamma rays. Each component is considered separately. The *TVL* for scattered and leakage photons (H_{Tot}) varies between 3 and 6 mm of lead depending on the maze length (McGinley, 2002), whereas the *TVL* for neutron capture gamma rays (H_{cg}) can be as much as 61 mm of lead (NCRP, 1984) depending on maze length. For this example, the shielding for the neutron capture gamma rays will suffice for the scattered- and leakage-radiation (H_{Tot}) components if it is assumed that the photon spectrum at the door is dominated by neutron capture gamma rays. Therefore, it is not necessary to calculate separately the shielding for the H_{Tot} contribution at the door.

In a situation like this, where the lead used to attenuate the neutron capture gamma rays is nearly transparent to the neutrons and the thickness of BPE needed for the neutrons weakly attenuates the neutron capture gamma rays, the following approach is straightforward and conservatively safe. Independently determine the material thickness for each radiation needed to achieve one-half of the shielding design goal.

The weekly neutron dose equivalent at the maze entrance was found to be (using Equation 2.21):

$$H_n = 765 \ \mu\text{Sv week}^{-1}$$

To reduce this neutron dose equivalent to $P/2 = 50 \ \mu\text{Sv week}^{-1}$, the number of *TVL*s required is:

$$n = \log\left(\frac{765}{50}\right) = 1.19$$

Using a *TVL* of 45 mm for BPE (Section 2.4), the required thickness for neutron shielding (with the additional *HVL*) is (1.19) (45 mm) = 53.6 mm of BPE.

The weekly contributions from the neutron capture gamma-ray dose equivalent, $H_{cg} = 65.3 \ \mu\text{Sv week}^{-1}$, is attenuated to a level of $50 \ \mu\text{Sv week}^{-1}$ with $n = \log(653 / 50) = 0.12$ *TVL*s. Using the *TVL* of 61 mm of lead (Section 2.4.3) for neutron capture gamma rays, the thickness is found to be 7.3 mm of lead.

Since 7.3 mm is >1 *TVL* for scattered and leakage photons, this amount of lead will suffice for both photon contributions, H_{cg} and H_T.

The total shielding for the maze door is ~7 mm of lead and 54 mm of BPE. The borated polyethylene (BPE) should be sandwiched between two layers of lead, each 3.5 mm thick (Section 2.4.3). However, this door design (weighing roughly 400 kg) may be sub-optimal, since the shielding contributions from the steel encasement of the door have been omitted. Methods for optimizing maze door design are considered by McGinley (2002).

IMRT modifications:

The higher leakage-radiation production with IMRT leads to a total dose equivalent at the maze door of:

$$H_w = H_{Tot} + H_{cg} + H_n$$
$$= (113 + 170 + 1{,}989)\ \mu Sv\ week^{-1} = 2{,}272\ \mu Sv\ week^{-1}$$

The H_{Tot} and H_{cg} contributions are 5 and 7.5 % of the total, respectively.

Door shielding for neutrons to $P/2$ = 50 μSv week^{-1} requires n = log (1,989 / 50) = 1.6 *TVL*s. The required thickness for neutron shielding is (1.6) (45 mm) = 72 mm of BPE. For the neutron capture gamma-ray contribution to be reduced to 50 μSv week^{-1}, the value for n = log (170 / 50) = 0.53 *TVL*. Using 61 mm of lead for the *TVL* yields a lead thickness of (0.53) (61 mm) = 32 mm.

Omitting consideration of the steel encasement, the door composition is 72 mm of BPE followed by 32 mm of lead. This is a heavy door weighing roughly 1,200 kg. The weight can be reduced by accounting for the steel encasement and the attenuation of neutron capture gamma rays provided by the BPE.

7.1.14 *Primary Barrier for Roof Location G*

The ceiling and roof for Sections 7.1.14 through 7.1.16 is shown in Figure 7.3.

In this example it is assumed that the roof top area is unoccupied open space and that there are no adjacent buildings toward which the primary beam points. The potential problem with these assumptions is that there may be no guarantee against future changes such as construction of an adjacent building. If the ceiling shielding is not sufficient, significant and costly problems may result from: (1) transmitted radiation to adjacent buildings with occupied levels higher than the roofline of the radiation therapy vault, and (2) atmospherically "reflected" photons and neutrons (skyshine) to ground and higher levels outside the vault. Skyshine radiation to the adjacent parking lot attendant or security guard may be unacceptable for a minimally shielded roof.

For example, the ceiling structure provides a primary barrier for Location G on the roof. The shielding material (ordinary concrete) is most economically poured in uniform thicknesses, without tapering. The roof is an area accessible by ladder only. It will occasionally have maintenance personnel working on the air conditioning units

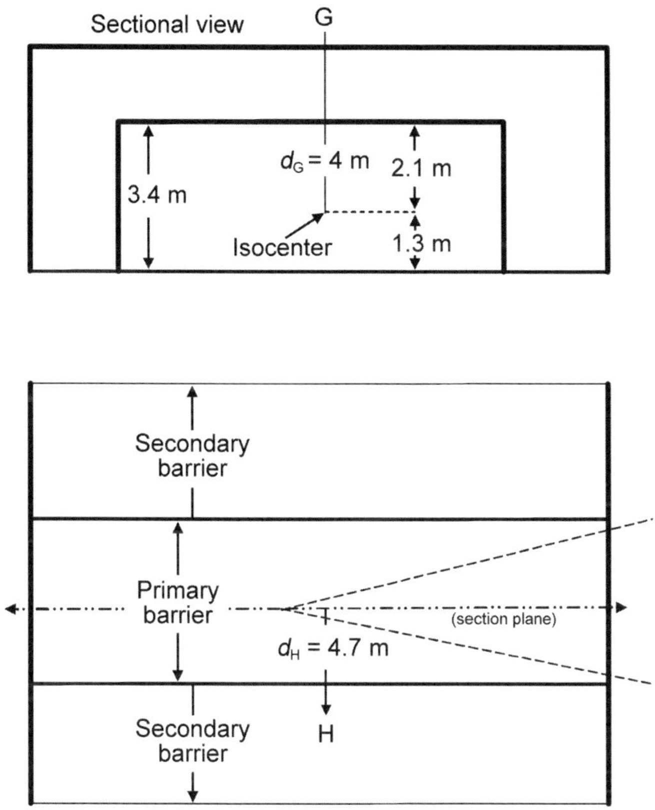

Fig. 7.3. The primary- and secondary-barrier geometry. Upper drawing is sectional view; lower drawing is floor plan.

and water chiller located (beyond the secondary barrier) ~15 m from the beam centerline. It is estimated that maintenance workers will spend an average of 2 h month^{-1} in the equipment area. The building supervisor has agreed to make a request with the radiation oncology physicist if more than 1 h is needed for a visit to the roof area. The calculated occupancy factor is (2) $(40 \times 4)^{-1}$ = 1/80. Therefore, the qualified expert has decided to use T = 1/40 as an occupancy factor.

Input data for Equation 2.1 are:

U = 0.25

P = 20×10^{-6} week^{-1}

T = 1/40

d_G = 4 m (isocenter to ceiling gap + barrier thickness + 0.3 m)
TVL_1(18 MV) = 45 cm and TVL_e(18 MV) = 43 cm
TVL_1(6 MV) = 37 cm and TVL_e(6 MV) = 33 cm

For the 18 MV beam, the primary-barrier thickness required is:

$$B(18\text{ MV}) = P\, d_G^2\, W^{-1}\, U^{-1}\, T^{-1}$$

$$= (20 \times 10^{-6})(4+1)^2(450)^{-1}(0.25)^{-1}(40)$$

$$= 1.78 \times 10^{-4}$$

$$n = 3.75$$

and

$$t(18\text{ MV}) = (45) + (2.75)(43) = 163\text{ cm}$$

For 6 MV:

$$B(6\text{ MV}) = 3.56 \times 10^{-4}$$

$$n = 3.45$$

and

$$t(6\text{ MV}) = (37) + (33)(2.45) = 118\text{ cm}$$

Since t(18 MV) exceeds t(6 MV) by 45 cm [~1 *TVL* (18 MV)], the larger thickness, 163 cm, qualifies as a minimum thickness to achieve the shielding design goal.

Note that the concrete primary barrier will be 3.4 m wide (matching the primary wall barrier width) and run the full length of the roof in the direction of the primary-beam swath.

The dose equivalent in-any-one-hour expected for the maximally exposed individual directly above the primary barrier (an unlikely situation) should be examined. Using Equations 3.8 and 3.14:

$$IDR(18\text{ MV}) = \dot{D}_o\, B(18\text{ MV})\, (d_G + 1)^{-2}$$
$$= (720)(1.8 \times 10^{-4})(5)^{-2}$$
$$= 5.2 \times 10^{-3}\text{ Sv h}^{-1}$$

$$R_w = IDR(18\text{ MV})\, W(18\text{ MV})\, U\, (\dot{D}_o)^{-1}$$
$$= (5.2 \times 10^{-3})(450)(0.25)(720)^{-1}$$
$$= 8.1 \times 10^{-4}\text{ Sv week}^{-1}$$

Similarly for 6 MV:

$$R_w(6\text{ MV}) = IDR\; W(6\text{ MV})\, U\, (\dot{D}_o)^{-1}$$
$$= B(6\text{ MV, 163 cm})\, W(6\text{ MV})\, U\, (d_G + 1)^{-2}$$
$$= 10^{-\left\{1 + \left[\frac{(163 - 37)}{33}\right]\right\}}(225)(0.25)(5)^{-2}$$
$$= 3.4 \times 10^{-5}\text{ Sv week}^{-1}$$

Thus:

$$R_h = \left(\frac{M}{40}\right) R_w(\text{total})$$
$$= (1.8)(40)^{-1}(8.1 \times 10^{4} + 3.4 \times 10^{-5})$$
$$= 38 \times 10^{-6}\text{ Sv} = 38\ \mu\text{Sv in-any-one-hour}$$

The result, 38 μSv, for an operational average use of both 6 and 18 MV x rays, exceeds the TADR limit of 20 μSv in-any-one-hour for the maximum dose equivalent. It is therefore recommended to add 1 *HVL*(18 MV) (0.301 × 43 = 12.9 cm) to the barrier thickness, bringing the total to 176 cm of concrete. The resulting value of R_h is <19 μSv in-any-one-hour due to spectral differences between the 6 and 18 MV beams, just under the TADR limit. However, the low occupancy (2 h or less per month), controlled access to the roof area, and low likelihood of anyone getting on top of the primary barrier, all suggest this level is safe and acceptable. Furthermore, if the same individual spent 2 h per month (0.5 per 40 h week), 12 months per year on top of the primary barrier under this worse case accelerator usage, the cumulative annual dose equivalent (24 h of exposure) would be 0.46 mSv which is less than the shielding design goal of 1 mSv y^{-1} for a member of the public. It is also noted that R_w(total) T, which is the time averaged dose equivalent to an individual at Location G per week of operation, has a value of <10.5 μSv $week^{-1}$ [$(81 \times 10^{-5} + 3.4 \times 10^{-5})(0.5/40)$] or <0.53 mSv y^{-1} (for a 50 week year).

7.1.15 *Secondary Barrier for Roof Location H*

Location H is taken as the closest point outside the primary barrier on a ray line from the isocenter (Figure 7.4).

First, consider the *patient-scattered radiation* from the isocenter to Location H, which is located ~30 degrees off the beam central axis from isocenter. The input data are similar to the data in Section 7.1.5 except:

$$T = 1/40$$

$$d_{sec} = d_H = d_L = 3.9 \text{ m}$$

For 18 MV patient-scattered radiation at 30 degrees, using Equation 2.7 and a use factor of 0.25, the barrier slant thickness is computed as follows:

$$B_{sca}(18 \text{ MV}) = P\,(a\,W\,U\,T)^{-1}\,d_{sca}^2\,d_{sec}^2\,(400)\,F^{-1}$$

$$= (20 \times 10^{-6})(3.9)^2\left[(2.53 \times 10^{-3})(450)(0.25)\left(\frac{1}{40}\right)\right]^{-1}$$

$$(400)(1{,}600)^{-1}$$

$$= 1.07 \times 10^{-2}$$

$$n = 1.97$$

and

$$t_{s,sca}(18 \text{ MV}) = (1.97)(32 \text{ cm}) = 63 \text{ cm}$$

See Table B.5a for the *TVL* data.

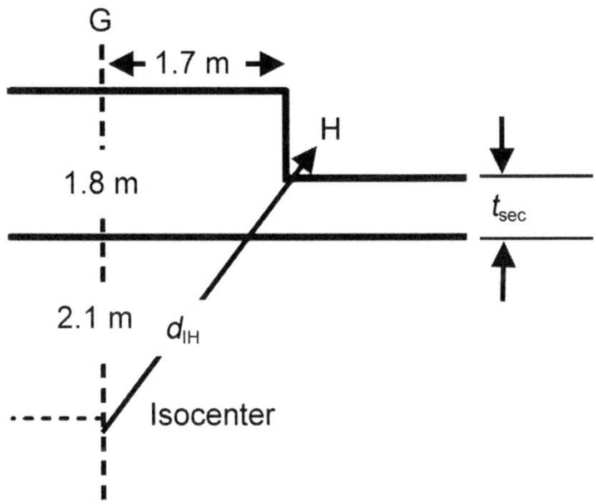

Fig. 7.4. Location H is taken as the closest point outside the primary barrier on a ray line from the isocenter (t_{sec} is estimated to be $\simeq$ 1 m, therefore $d_{IH} \simeq 3.9$ m).

For 6 MV scattered x rays:

$$B_{sca}(6\text{ MV})$$

$$= (20 \times 10^{-6})(3.9)^2\left[(2.77 \times 10^{-3})(225)(0.25)\left(\frac{1}{40}\right)\right]^{-1}(0.25)$$

$$= 1.95 \times 10^{-2}$$

$$n = 1.71$$

and

$$t_{s,sca}(6\text{ MV}) = (1.71)(26\text{ cm}) = 44.5\text{ cm}$$

The 18 MV *leakage radiation* requires a barrier slant thickness, according to Equation 2.2, of:

$$B_L(18\text{ MV}) = (10^3)(20 \times 10^{-6})(3.9)^2(450)^{-1}(40)$$

$$= 2.7 \times 10^{-2}$$

$$n = 1.57$$

and

$$t_{s,L}(18\text{ MV}) = (36\text{ cm}) + (0.57)(34\text{ cm}) = 55.4\text{ cm}$$

For 6 MV leakage radiation:

$$B_L(6\text{ MV}) = (10^3)(20 \times 10^{-6})(3.9)^2(225)^{-1}(40) = 5.4 \times 10^{-2}$$

$$n = 1.27$$

and

$$t_{s,L}(6\text{ MV}) = (34\text{ cm}) + (0.27)(29\text{ cm}) = 41.8\text{ cm}$$

The highest value for the independently calculated slant thickness for Location H is 63 cm, coming from 18 MV patient-scattered radiation at 30 degrees and the next highest value is 55.4 cm from 18 MV leakage radiation. Adding a 18 MV *HVL* for scattered radiation [(32 cm) (0.301) = 9.6 cm] to the highest value yields 72.6 cm. The next highest value for slant thickness is 44.5 cm from 6 MV scattered radiation, which is ~28 cm < 72.6 cm. Since 1 *TVL* for 6 MV scattered radiation is 26 cm (<28 cm), additional concrete is not needed. Taking the barrier slant thickness for secondary radiation ($t_{s,sec}$) as 73 cm, the barrier adequacy for Location H can be verified as follows:

$$H_L(18\text{ MV}) = 10^{-\left\{1+\left[\frac{(73-36)}{34}\right]\right\}}(10^{-3})\ W_L\ T\ (d_L)^{-2}$$

$$= 6\ \mu\text{Sv week}^{-1}$$

$$H_L(6\text{ MV}) = 10^{-\left\{1+\left[\frac{(73-34)}{29}\right]\right\}}(10^{-3})\ W_L\ T\ (d_L)^{-2}$$

$$= 1.7\ \mu\text{Sv week}^{-1}$$

$$H_{sca}(18\text{ MV}) = 10^{-\left(\frac{73}{32}\right)}\ a\ (d_{sec})^{-2}\ (4)\ (W\ U\ T) = 9.8\ \mu\text{Sv week}^{-1}$$

$$H_{sca}(6\text{ MV}) = 10^{-\left(\frac{73}{26}\right)}\ a\ (d_{sec})^{-2}\ (4)\ (W\ U\ T) = 1.6\ \mu\text{Sv week}^{-1}$$

The total of all contributions is 19 μSv week^{-1}, which is acceptable.

The thickness of concrete for the ceiling secondary barrier (t_{sec}) is:

$$t_{sec} = \cos(30°)\ t_{s,sec} = 63.2 \text{ cm}$$

IMRT modifications:

The increased leakage radiation due to IMRT procedures results in the larger computed slant thicknesses for leakage radiation:

$$B_L(18 \text{ MV}) = (10^3)\ P\ d_L^2\ [W_L(18 \text{ MV})\ T]^{-1}$$

$$= 1.04 \times 10^{-2}$$

$$n = 1.98$$

and

$$t_{s,L}(18 \text{ MV}) = (36 \text{ cm}) + (0.98)(34 \text{ cm}) = 69.3 \text{ cm}$$

For 6 MV leakage radiation:

$$t_{s,L}(6 \text{ MV}) = (34 \text{ cm}) + (0.89)(29 \text{ cm}) = 59.8 \text{ cm}$$

With IMRT, the 18 MV leakage radiation requires the highest value for the slant thickness (69.3 cm), followed by the 18 MV patient-scattered-radiation thickness of 63 cm. Adding 1 *HVL* (10.8 cm) for 18 MV leakage radiation to 69.3 cm gives a combined slant thickness of 80.1 cm, which is ~7 cm more than without IMRT (73 cm). The secondary-barrier thickness is: t_{sec} = [cos (30 degrees)] (80.1 cm) = 69.4 cm of ordinary concrete.

Verifying the total of all attenuated contributions at Location H, the leading two contributions are found to be 9.7 μSv week^{-1} from 18 MV leakage radiation and 5.9 μSv week^{-1} from 18 MV patient-scattered radiation. All four contributions total 20.5 μSv week^{-1}, which is just over the shielding design goal of 20 μSv week^{-1}, suggesting that another 11 cm of concrete should be added, bringing the slant thickness to 91 cm and the barrier thickness to t_{sec} = 79 cm. This lowers the weekly total dose equivalent to 9.3 μSv, which is acceptable.

7.1.16 *Time Averaged Dose-Equivalent Rate for the Secondary Barrier at Location H with Intensity Modulated Radiation Therapy*

Following the same methods as in Section 7.1.6 (using the final barrier slant thickness of 91 cm):

$$IDR_{\mathrm{L}}(18\ \mathrm{MV}) = \dot{D}_{\mathrm{o}}\ (10^{-3})\ B_{\mathrm{L}}(18\ \mathrm{MV}, 91\ \mathrm{cm})\ (d_{\mathrm{L}})^{-2}$$

$$= 1.14 \times 10^{-4}\ \mathrm{Sv\ h^{-1}}$$

$$IDR_{\mathrm{L}}(6\ \mathrm{MV}) = 0.51 \times 10^{-4}\ \mathrm{Sv\ h^{-1}}$$

$$IDR_{\mathrm{ps}}(18\ \mathrm{MV}) = \dot{D}_{\mathrm{o}}\ a\ F\ (400)^{-1}\ B_{\mathrm{sca}}(18\ \mathrm{MV}, 30°)\ (d_{\mathrm{sec}})^{-2}$$

$$= 6.86 \times 10^{-4}\ \mathrm{Sv\ h^{-1}}$$

$$IDR_{\mathrm{ps}}(6\ \mathrm{MV}) = 1.66 \times 10^{-4}\ \mathrm{Sv\ h^{-1}}$$

$$R_{\mathrm{w}}(18\ \mathrm{MV}) = [IDR_{\mathrm{L}}\ W_{\mathrm{L}}\ (\dot{D}_{\mathrm{o}})^{-1}] + [IDR_{\mathrm{ps}}\ W_{\mathrm{ps}}\ U_{\mathrm{ps}}\ (\dot{D}_{\mathrm{o}})^{-1}]$$

$$= 2.93 \times 10^{-4}\ \mathrm{Sv\ week^{-1}}$$

$$R_{\mathrm{w}}(6\ \mathrm{MV}) = 0.8 \times 10^{-4}\ \mathrm{Sv\ week^{-1}}$$

The dose equivalent to an individual in-any-one-hour comes from treating 10 patients in-any-one-hour. The operational total TADR for a week (R_{w}) is [summing R_{w}(6 MV) and R_{w}(18 MV)] 3.7×10^{-4} Sv week^{-1}. This yields a dose equivalent in-any-one-hour of $R_{\mathrm{h}} = (M/40)\, R_{\mathrm{w}} = 16.7$ μSv, which is below the TADR limit of 20 μSv.

7.1.17 *Maze Barrier Thickness*

The concrete thickness of the maze barrier can be determined by considering the photon leakage radiation transmission and, for high energy machines, the transmission of fast photoneutrons to the location of the maze door.

The 6 and 18 MV x-ray leakage radiation dose equivalents at the door are estimated using Equation 2.8. For the IMRT situation, the data are:

W_L(18 MV) = 1,170 Gy week^{-1}
W_L(6 MV) = 945 Gy week^{-1}
TVL_1(18 MV) = 36 cm and TVL_e(18 MV) = 34 cm for leakage radiation in concrete
TVL_1(6 MV) = 34 cm and TVL_e(6 MV) = 29 cm for leakage radiation
T = 1
U = 1
P = 0.1 × 10^{-3} Sv week^{-1}
d_{hd} = 7.7 m from isocenter to door

For 18 MV:

$$B(18\ \text{MV}) = P\,(d_{\text{ld}})^2(10^3)(W_{\text{L}}\ T)^{-1}$$

$$= (10^{-4})\ (7.7)^2\ (10^3)\ (1{,}170)^{-1}$$

$$= 5.07 \times 10^{-3}$$

$$n = 2.3$$

and the barrier slant thickness is:

$$t_{\text{s,L}}(18\ \text{MV}) = (36\ \text{cm}) + (1.3)(34\ \text{cm}) = 80.2\ \text{cm}.$$

Similarly, for 6 MV:

$$t_{\text{s,L}}(6\ \text{MV}) = 69\ \text{cm}$$

To combine these slant thicknesses, since they differ by <1 *TVL*, 1 *HVL* is added to the higher value. The total slant thickness for leakage photons is:

$$t_{\text{s,L}}(\text{total}) = (80.2\ \text{cm}) + (0.301)(36\ \text{cm}) = 91\ \text{cm}$$

Next consider the transmission of fast photoneutrons through the concrete maze barrier. It has been suggested (NCRP, 1977; 2003) that the barrier thickness of concrete required for primary photons can suffice for the attenuation of photoneutron dose equivalent to an acceptable level.

The situation to be examined here is for leakage x-ray radiation for which *TVL*s may be less than for primary *TVL*s. After buildup in the first *TVL*, Figure A.2 gives dose-equivalent *TVL*s of ~23 to 27 cm for neutrons with average energy in the range of 0.35 to 2.7 MeV. This range of TVL values is below the range of photon leakage-radiation TVL_e values (33 to 36 cm) listed in Table B.7 for high-energy photons in ordinary concrete. The fast neutron fluence at the door in the absence of the maze barrier is estimated for the first term in Equation 2.16. Taking $\beta = 1$ and obtaining $Q_n = 1.2 \times 10^{12}$ n Gy^{-1} from Table B.9 for the neutron source strength per photon-gray at isocenter, the direct-neutron fluence at the door is:

$$\varphi_n = Q_n\,(4\pi\, d_{1d}^2)^{-1} = (1.2 \times 10^{12}\ \text{n Gy}^{-1})(4\pi)^{-1}(770\ \text{cm})^{-2}$$

$$= 1.6 \times 10^5\ \text{n cm}^{-2}\ \text{Gy}^{-1}$$

From Figure A.2, it is found that 91 cm of ordinary concrete (214 g cm^{-2}) provides a neutron dose-equivalent of $H_{ns} \approx 10^{-13}$ Sv cm^2 n^{-1} for the direct neutrons from the accelerator head [*i.e.*, $\bar{E}_n$ on the order of 1 MeV (NCRP, 1984)]. For the IMRT workload for 18 MV x rays the neutron dose equivalent at the door is:

$$H_n(\text{door}) = H_{ns}\, W_L\, \varphi_n \simeq 19\ \mu\text{Sv week}^{-1}$$

This is lower than the combined photon contribution, 67 μSv $week^{-1}$, at the door with a 91 cm concrete barrier slant thickness.

In view of the uncertainty in the above estimation and of all the other significant contributions at the maze door, the qualified expert has recommended that the barrier slant thickness be increased to ~125 cm. Thus, the barrier thickness (t), where ray-lines enter the barrier on the way to the door from the accelerator, is t = 125 cm [cos (40 degrees)] = 96 cm.

If IMRT was not an option for this installation, then the maze barrier thickness could be decreased.

7.2 Robotic Arm Stereotactic-Radiosurgery Room

This example illustrates a shielding design for a robotically mounted compact 6 MV x-ray source used for stereotactic radiosurgery (SRS). The primary beam can be pointed at most of the treatment room walls. However, the primary beam does not reach the ceiling for the size of the room designated in this example. The ceiling or roof barrier is the only secondary barrier for this kind of treatment unit. Since the unit is nonisocentric and uses hundreds of small (typically 1.5 cm diameter) circular fields to irradiate the target volume, the value for C_I (for use in Equation 3.3) is relatively large. Analysis of clinical data (Rodgers, 2005) yields an average C_I value of 15. As a consequence of this C_I value, the leakage radiation is relatively high and may contribute to the primary-barrier thickness requirements. A robotic manipulator does not need to have an isocenter of motion. The system design attempts to keep the accelerator at roughly a fixed distance from the treatment volume during the course of treatment delivery. The pseudo-isocentric distance is taken as 80 cm and the absorbed dose per monitor unit and absorbed-dose rate are specified as 1 cGy MU^{-1} and 4 Gy min^{-1}, respectively, at 80 cm source-to-phantom distance.

A workload W is determined given that treatments are delivered 5 d week^{-1} with up to eight (cranial and extra-cranial) patients treated per day. The mean absorbed dose per treatment (an average of single and multiple fractionated cases) is 12.5 Gy.

The recommended value for the use factor (0.05) has been determined from clinically measured data (Rodgers, 2005). It is easily verified that the patient-scattered dose-equivalent contributions are negligible compared to the contributions from leakage radiation.

The floor plan for the room is shown in Figure 7.5, in which the distances (d) from the pseudo-isocenter to the points of interest are indicated. All barriers in this example are to be made from ordinary concrete.

The required barrier thickness at Location A will be computed to shield this area to the shielding design goal for an uncontrolled area, since the qualified expert anticipates that visitors will be in this area for long periods. Equation 2.1 for primary-beam barriers is adaptable to this situation, where:

P = 20×10^{-6} Sv week^{-1}
d_A = distance from the isocenter to 0.3 m beyond the barrier = 4.9 m
W = workload for a 5 d week = 500 Gy week^{-1} ($8 \times 5 \times 12.5$) for the primary-beam x rays at a nominal distance of 0.80 m from the x-ray target. Since W (at 0.80 m) (0.80 m)^2 = W(at 1 m) (1 m)^2, the weekly workload at 1 m is W (at 1 m) = 320 Gy week^{-1}.
U = use factor = 0.05
T = occupancy factor for the control station = 1

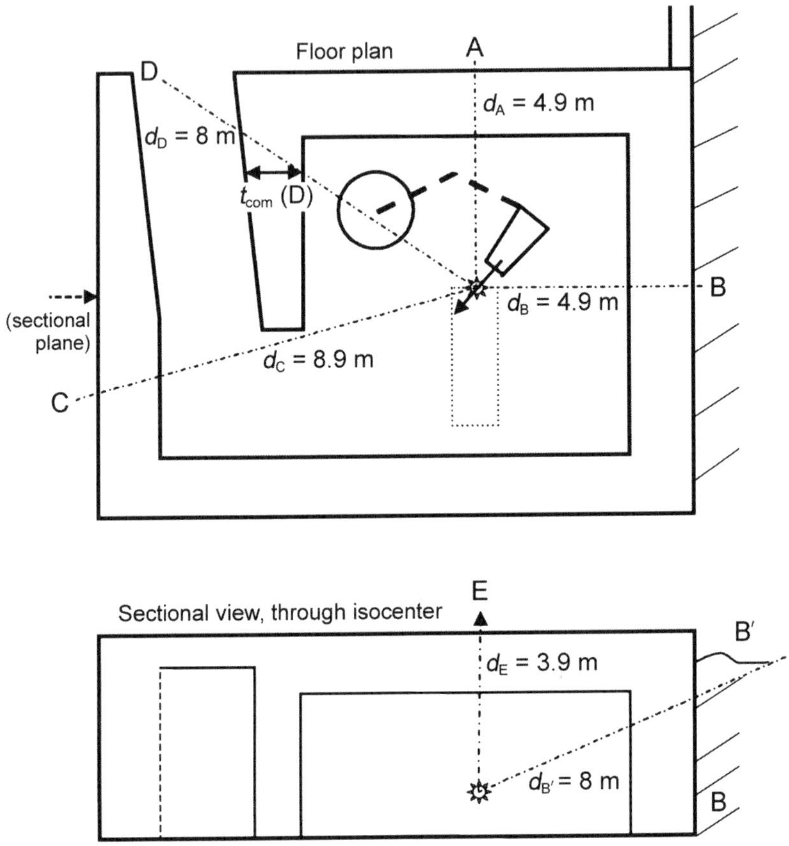

Fig. 7.5. Robotic arm room layout. Upper drawing is the floor plan; lower drawing is a sectional view.

$$B_{pri} = P\,(d_A + 0.8)^2\ [W(\text{at 1 m})\ U\ T]^{-1}$$

$$= \frac{(20 \times 10^{-6})(4.9 + 0.8)^2}{(320)(0.05)(1)} = 4.06 \times 10^{-5}$$

The required number of *TVL*s to produce this attenuation is determined from Equation 2.2:

$$n = \log\left(\frac{1}{4.06 \times 10^{-5}}\right) = 4.39$$

The TVL_e for a broad beam of 6 MV x rays is 33 cm in ordinary concrete (Table B.2) and will be used as a conservatively safe value for both TVL_1 and TVL_e in this small beam application. Therefore, the barrier thickness computed for primary radiation at Location A is:

$$t_{pri}(A) = (4.39)(33 \text{ cm}) = 145 \text{ cm}$$

The leakage-radiation workload at 1 m from the x-ray target (in the absence of head shielding) is:

$$W_L = C_I \, W(\text{at } 0.80 \text{ m}) \, (0.64 \text{ cGy MU}^{-1}) = C_I \, W(\text{at } 1 \text{ m})$$

$$= (15 \text{ MU cGy}^{-1})(500 \text{ Gy week}^{-1})(0.64 \text{ cGy MU}^{-1})$$

$$= 4.8 \times 10^3 \text{ Gy week}^{-1}$$

Since there is no measured *TVL* for leakage radiation, the primary radiation TVL_e of 33 cm will be used as a conservatively safe value.

From Equation 2.8, the barrier thickness requirement for leakage radiation is:

$$B_L = (10^3) \, P \, d_A^2 \, W_L^{-1} \, T^{-1} = 1 \times 10^{-4}$$

$$n = 4$$

and

$$t_L(A) = (4)(33 \text{ cm}) = 122 \text{ cm}$$

The required concrete thicknesses for primary-beam and leakage radiations differ by 23 cm, which is <1 *TVL*. Therefore, the required barrier thickness at Location A [t(A)] is 145 cm plus 1 *HVL* (9.9 cm), which is 155 cm.

Note, that if the shielding design goal were for a controlled area value of 100 μSv week^{-1}, the concrete barrier thickness would be 132 cm.

7.2.1 *Time Averaged Dose-Equivalent Rate Considerations for Location A*

The maximum dose equivalent in-any-one-hour and the average dose equivalent per week are determined by the TADR quantities

R_h and R_w (Equations 3.6 and 3.8). The absorbed-dose rate at 80 cm ($\dot{D}_o$) is 4 Gy min^{-1} or 240 Gy h^{-1}. Hence, the absorbed-dose rate at 1 m is 154 Gy h^{-1}. Therefore, the instantaneous dose-equivalent rate from the primary beam (IDR_{pri}) on the far side of the barrier protecting Location A is:

$$\begin{aligned} IDR_{pri} &= \dot{D}_o\, B\, (d_A + 0.8)^{-2} \\ &= (154 \text{ Gy h}^{-1})\, 10^{-\left(\frac{155}{33}\right)}\, (4.9 \text{ m} + 0.8 \text{ m})^{-2} \\ &= 95\ \mu\text{Sv h}^{-1} \end{aligned}$$

The primary beam produces a weekly TADR at Location A of:

$$\begin{aligned} R_w(\text{pri}) &= IDR_{pri}\, W_{pri}\, U_{pri}\, (\dot{D}_o)^{-1} \\ &= (95\ \mu\text{Sv h}^{-1})(320 \text{ Gy week}^{-1})(0.05)(154 \text{ Gy h}^{-1})^{-1} \\ &= 9.9\ \mu\text{Sv week}^{-1} \end{aligned}$$

The leakage-radiation contributions need to be considered as well. The instantaneous dose-equivalent rate from leakage radiation (IDR_L) at Location A is:

$$\begin{aligned} IDR_L &= (10^{-3})\, \dot{D}_o\, B_L\, d_A^{-2} \\ &= (10^{-3})(154 \text{ Gy h}^{-1})\, 10^{-\left(\frac{155}{33}\right)}\, (4.9 \text{ m})^{-2} \\ &= 0.129\ \mu\text{Sv h}^{-1} \end{aligned}$$

Thus, the weekly TADR from leakage radiation is:

$$\begin{aligned} R_w(\text{L}) &= IDR_L\, W_L\, (\dot{D}_o)^{-1} \\ &= (1.29 \times 10^{-7} \text{ Sv h}^{-1})(4.8 \times 10^3 \text{ Gy week}^{-1})(1.54 \times 10^2 \text{ Gy h}^{-1})^{-1} \\ &= 4\ \mu\text{Sv week}^{-1} \end{aligned}$$

The sum R_w(pri) + R_w(L) = 13.9 μSv week^{-1}, which is below the shielding design goal of 20 μSv week^{-1}.

To compute the maximum dose equivalent at Location A in-any-one-hour, the value of M (Equation 3.14) is needed, where M is the ratio of the maximum number of treatments in an hour to the average. Workload (W) considerations are based on an average of one treatment per hour and it is assumed that the maximum possible is two treatments in an hour. Therefore, $M = 2$ and the maximum dose equivalent in-any-one-hour is:

$$R_h = \left(\frac{M}{40}\right) R_w$$

$$= \left(\frac{2}{40}\right) 13.9 \ \mu\text{Sv}$$

$$= 0.7 \ \mu\text{Sv in-any-one-hour}$$

This is well below the TADR limit of 20 μSv in-any-one-hour.

7.2.2 *Barrier for Location B*

Location B is below grade, so the occupancy factor is zero. The wall contributes to the support of the roof barrier. Leakage and primary radiations contribute to the dose equivalent at Location B′ (0.3 m above grade). The manufacturer of the accelerator states that the primary beam may point as high as 22 degrees above the horizontal. The ray line from isocenter to Location B′ is 22 degrees above the horizontal and is located a distance of ~8 m ($d_{B'}$) from the isocenter. The path length through the ground is ~1.5 m. This is an uncontrolled area. Since this area is not an office and only transient occupancy is ever expected, the occupancy factor will be taken as 1/20.

The primary and leakage-radiation barrier slant thicknesses $t_{s,pri}(B')$ and $t_{s,L}(B')$ are determined as follows:

$$B_{pri} = P\,(d_{B'} + 0.8)^2\,(W\,U\,T)^{-1}$$

$$= (20 \times 10^{-6}\ \text{Sv week}^{-1})(8.8\ \text{m})^2[(320\ \text{Gy week}^{-1})(0.05)(0.05)]^{-1}$$

$$= 1.94 \times 10^{-3}$$

$$n = 2.71$$

and

$$t_{s,pri}(B') = (2.71)(33\text{ cm}) = 89.4\text{ cm}.$$

while

$$B_L = (10^3)\,P\,d_{B'}^2\,(W_L\,T)^{-1}$$
$$= (10^3)(20 \times 10^{-6})(8)^2[(4{,}800)(0.05)]^{-1}$$
$$= 5.33 \times 10^{-3}$$

$$n = 2.27$$

and

$$t_{s,L}(B') = (2.27)(33\text{ cm}) = 75\text{ cm}.$$

Since the primary and leakage-radiation barrier slant thicknesses differ by <1 *TVL*, the required slant thickness at Location B′ is t (B′) = 99 cm of concrete or the equivalent of 3 *TVL*s (99 / 33).

The 1.5 m of earth provides some shielding benefit. Since the mass density of compact earth is typically taken as 1.5 g cm^{-3} (NCRP, 1976), the mass density of earth relative to concrete is 1.5 / 2.35 = 0.64. This yields an estimated *TVL* of 52 cm [(33 cm) $(0.64)^{-1}$] for earth in this situation. Thus, 1.5 m of earth provides ~2.88 *TVL*s. The remaining 0.12 *TVL* (3 – 2.88) must come from the concrete barrier. A minimum of 4 cm of concrete (barrier slant thickness) is required.

7.2.3 *Shielding for Location C*

Location C (Figure 7.5) and the area adjacent to this barrier is a mechanical support room housing equipment for the accelerator and robot. It has controlled access and is restricted to badged workers when radiation is being produced. The shielding design goal is taken as 0.1 mSv $week^{-1}$. The qualified expert has determined that an occupancy factor of $T = 1/8$, corresponding to 1 h d^{-1} access while radiation is being produced, is appropriate for this area. The angle of incidence to the normal is ~15 degrees.

The primary and leakage-radiation shielding requirements are:

$$B_{pri} = P\,(d_C + 0.8)^2\,(W\,U\,T)^{-1}$$
$$= (100 \times 10^{-6})(8.9 + 0.8)^2[(320)(0.05)(0.125)]^{-1}$$
$$= 4.7 \times 10^{-3}$$

$$n = 2.33$$

and

$$t_{s,pri}(C) = (2.33)(33\text{ cm}) = 77\text{ cm.}$$

while

$$B_L = (10^3)\,P\,d_C^2\,(W_L\,T)^{-1}$$
$$= (10^3)(100 \times 10^{-6})(8.9)^2[(4{,}800)(0.125)]^{-1}$$
$$= 1.32 \times 10^{-2}$$

$$n = 1.88$$

and

$$t_{s,L}(C) = 62\text{ cm.}$$

Adding 1 *HVL* to the primary-beam barrier slant thickness $t_{s,pri}(C)$ yields a combined (primary plus leakage) slant thickness [$t_{s,com}(C)$] of 87 cm of concrete. Since the angle of incidence with respect to the normal is ~15 degrees, the barrier thickness required is $t(C) = \cos(15\text{ degrees})\, t_{s,com}(C) = 84.4$ cm.

7.2.4 *Determination of the Maze Barrier Thickness (Location D)*

In this example, it was decided that the maze barrier must adequately shield the room doorway and beyond. The door shielding, if any, does not contribute to this consideration. In a separate calculation, room and patient-scattered radiation contributions will be added to the maze barrier transmitted dose equivalent at the door to determine the door-shielding requirement. Location D will be treated like the control station with the shielding design goal for an uncontrolled area and full occupancy assumed. The angle of incidence is 32 degrees.

The primary and leakage radiations at Location D determine a barrier thickness as follows:

$$B_{pri} = P\,(d_D + 0.8)^2\,(W\,U\,T)^{-1}$$

$$= (20 \times 10^{-6}\ \text{Sv week}^{-1})(8.8\ \text{m})^2[(320\ \text{Gy week}^{-1})(0.05)(1)]^{-1}$$

$$= 9.68 \times 10^{-5}$$

$$n = 4.01$$

and

$$t_{s,pri}(\text{D}) = (4.01)(33\ \text{cm}) = 132.3\ \text{cm}.$$

while

$$B_L = (10^3)\,P\,d_D^2\,(W_L\,T)^{-1}$$

$$= (10^3)(20 \times 10^{-6})(8)^2[(4{,}800)(1)]^{-1}$$

$$= 2.67 \times 10^{-4}$$

$$n = 3.57$$

and

$$t_{s,L}(\text{D}) = (3.57)(33\ \text{cm}) = 118\ \text{cm}.$$

Therefore, the needed barrier slant thickness $t_{s,com}$(D) = 132.3 + 9.9 cm = 142.2 cm. Since the angle of incidence is 32 degrees, the barrier thickness, where the ray line passes through the barrier, is t_{com}(D) = cos (32 degrees) (142.2) = 121 cm.

7.2.5 *Shielding at Location E (Roof)*

The primary beam does not contribute to this location. It is an uncontrolled area with limited access by virtue of a fence around the perimeter of the ceiling barrier. However, the qualified expert's conservatively safe approach will use an occupancy factor $T = 1/20 = 0.05$. The angle of incidence is taken as 90 degrees (zero degree obliquity).

The leakage radiation requires a concrete thickness of 95.7 cm, determined as follows:

$$B_L = (10^3)\ P\ d_E^2\ (W_L\ T)^{-1}$$

$$= (10^3)(20 \times 10^{-6})(3.9)^2[(4{,}800)(0.05)]^{-1}$$

$$= 1.27 \times 10^{-3}$$

$$n = 2.90$$

and

$$t_L(\mathrm{E}) = t(\mathrm{E}) = (2.9)(33\ \mathrm{cm}) = 95.7\ \mathrm{cm}$$

Appendix A

Supporting Data (Figures)

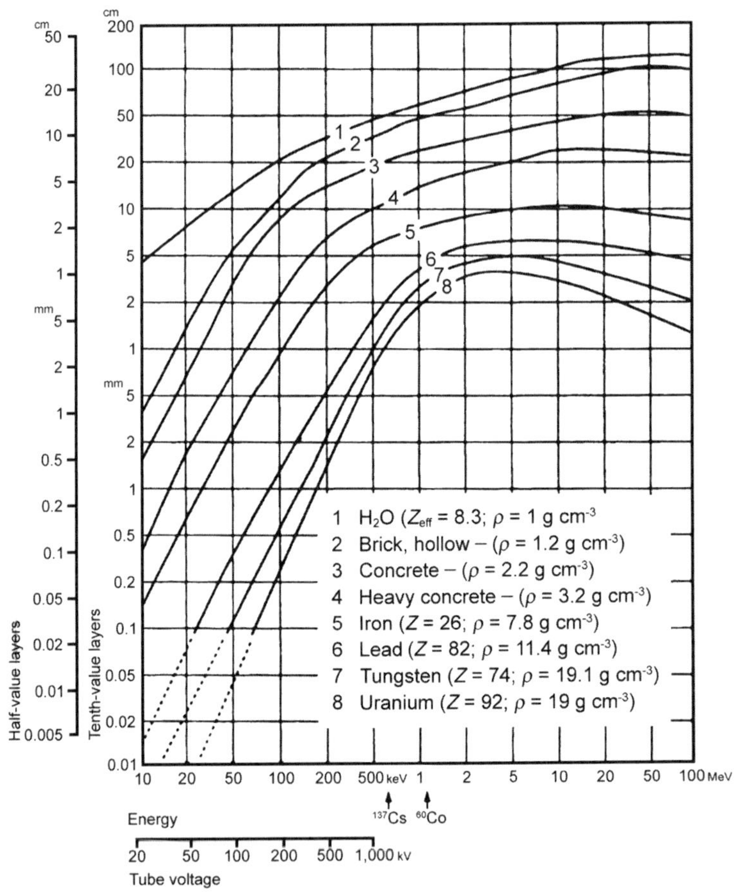

Fig. A.1a. Average *HVL*s and *TVL*s (equilibrium) of shielding materials (broad beams) (NBS, 1982; Wachsmann and Drexler, 1975). For example, an energy of 10 MeV gives a *TVL* in concrete (Curve 3) of ~44 cm and a *HVL* of ~13 cm. Note that these values will be less for concrete of density 2.35 g cm^{-3} by ~0.94 (*i.e.*, inversely proportional to the densities, 2.2/2.35).

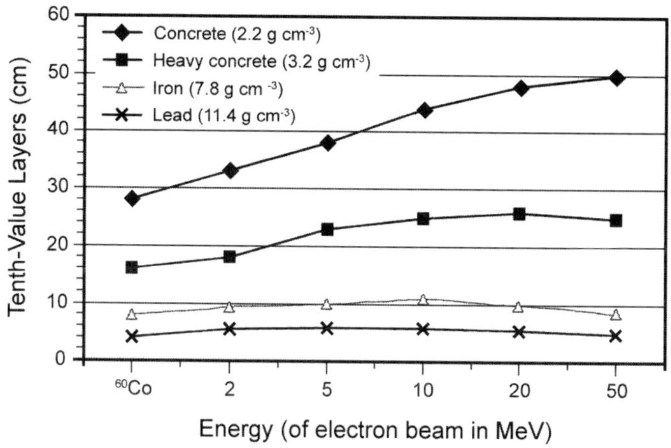

Fig. A.1b. Primary *TVL*s for materials (expanded from Figure A.1a).

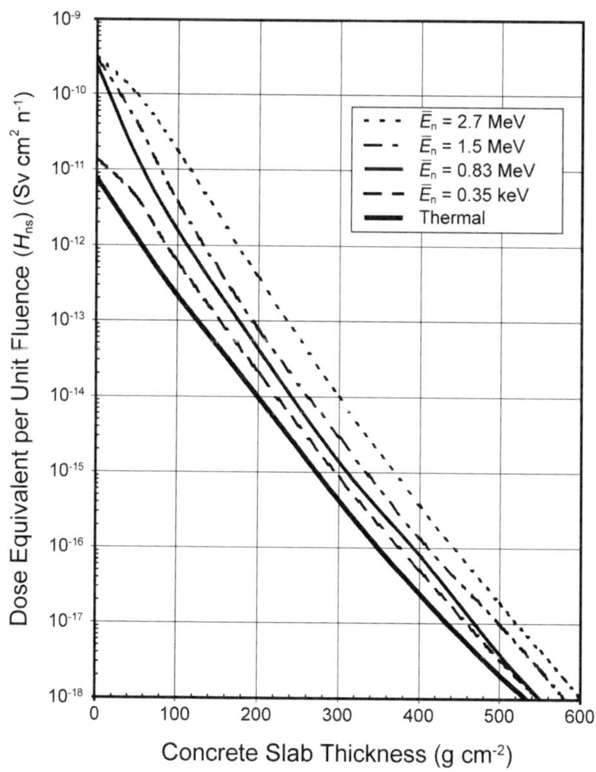

Fig. A.2. Calculated dose equivalent (including neutron capture gamma-ray contribution) transmitted per unit fluence of neutrons with average energy $\bar{E}_n$ incident normally on slabs of ordinary concrete (H_{ns}) (Sv cm^2 n^{-1}) (adapted from NCRP, 1977).

Appendix B

Supporting Data (Tables)

TABLE B.1—*Suggested occupancy factors*[a] *(for use as a guide in planning shielding when other sources of occupancy data are not available).*

Location	Occupancy Factor (T)
Full occupancy areas (areas occupied full-time by an individual), *e.g.*, administrative or clerical offices; treatment planning areas, treatment control rooms, nurse stations, receptionist areas, attended waiting rooms, occupied space in nearby building	1
Adjacent treatment room, patient examination room adjacent to shielded vault	1/2
Corridors, employee lounges, staff rest rooms	1/5
Treatment vault doors[b]	1/8
Public toilets, unattended vending rooms, storage areas, outdoor areas with seating, unattended waiting rooms, patient holding areas, attics, janitors' closets	1/20
Outdoor areas with only transient pedestrian or vehicular traffic, unattended parking lots, vehicular drop off areas (unattended), stairways, unattended elevators	1/40

[a]When using a low occupancy factor for a room immediately adjacent to a therapy treatment vault, care *shall* be taken to also consider the areas further removed from the treatment room. The adjacent room may have a significantly higher occupancy factor and may therefore be more important in shielding design despite the larger distances involved.

[b]The occupancy factor for the area just outside a treatment vault door can often be assumed to be lower than the occupancy factor for the work space from which it opens.

TABLE B.2—*Primary-barrier TVLs for ordinary concrete (2.35 g cm^{-3}), steel (7.87 g cm^{-3}), and lead (11.35 g cm^{-3}) (suggested values in centimeters).*[a]

Endpoint Energy (MV)[b]	Material	TVL_1 (cm)	TVL_e (cm)
4	Concrete	35	30
	Steel	9.9	9.9
	Lead	5.7	5.7
6	Concrete	37	33
	Steel	10	10
	Lead	5.7	5.7
10	Concrete	41	37
	Steel	11	11
	Lead	5.7	5.7
15	Concrete	44	41
	Steel	11	11
	Lead	5.7	5.7
18	Concrete	45	43
	Steel	11	11
	Lead	5.7	5.7
20	Concrete	46	44
	Steel	11	11
	Lead	5.7	5.7
25	Concrete	49	46
	Steel	11	11
	Lead	5.7	5.7
30	Concrete	51	49
	Steel	11	11
	Lead	5.7	5.7
Co-60	Concrete	21	21
	Steel	7.0	7.0
	Lead	4.0	4.0

[a]Concrete values are based on a conservatively safe adaptation from Nelson and LaRiviere (1984) with extrapolation to 4 MV, and use of Kirn and Kennedy (1954) for 30 MV. Lead and steel *TVLs* are conservatively safe values adapted from NCRP Report No. 49 (NCRP, 1976) and Wachsmann and Drexler (1975).

[b]Endpoint energy based on values from Cohen (1972).

TABLE B.3—*Properties of various shielding materials (adapted from Profio, 1979).*

	Ordinary Concrete	Heavy Concrete	Lead	Iron	Polyethylene
Density (g cm^{-3})	2.2 – 2.4	3.7 – 4.8	11.35	7.87	0.95
Effective atomic number	11	~26	82	26	5.5
Hydrogen concentration × 10^{22} (atoms cm^{-3})	0.8 – 2.4	0.8 – 2.4	0	0	8
Thermal neutron activation	Small	Large	—[a]	Moderate	Nil
Relative cost	$$	$$$$	$$$	$$	$$$

[a]The amount of thermal neutron activation depends primarily on the impurities in the lead.

TABLE B.4—*Scatter fractions (a) at 1 m from a human-size phantom, target-to-phantom distance of 1 m, and field size of 400 cm² (McGinley, 2002; Taylor et al., 1999).*

Angle (degrees)	Scatter Fraction (*a*)			
	6 MV	10 MV	18 MV	24 MV
10	1.04×10^{-2}	1.66×10^{-2}	1.42×10^{-2}	1.78×10^{-2}
20	6.73×10^{-3}	5.79×10^{-3}	5.39×10^{-3}	6.32×10^{-3}
30	2.77×10^{-3}	3.18×10^{-3}	2.53×10^{-3}	2.74×10^{-3}
45	1.39×10^{-3}	1.35×10^{-3}	8.64×10^{-4}	8.30×10^{-4}
60	8.24×10^{-4}	7.46×10^{-4}	4.24×10^{-4}	3.86×10^{-4}
90	4.26×10^{-4}	3.81×10^{-4}	1.89×10^{-4}	1.74×10^{-4}
135	3.00×10^{-4}	3.02×10^{-4}	1.24×10^{-4}	1.20×10^{-4}
150	2.87×10^{-4}	2.74×10^{-4}	1.20×10^{-4}	1.13×10^{-4}

TABLE B.5a—*TVLs in concrete (centimeters) for patient-scattered radiation at various scatter angles, based on Figures 10 and 15 in NCRP Report No. 49 (NCRP, 1976). Values are valid for shielding design purposes and are conservatively safe in nature.*[a]

Scatter Angle (degrees)	*TVL* (cm)							
	Co-60	4 MV	6 MV	10 MV	15 MV	18 MV	20 MV	24 MV
15	22	30	34	39	42	44	46	49
30	21	25	26	28	31	32	33	36
45	20	22	23	25	26	27	27	29
60	19	21	21	22	23	23	24	24
90	15	17	17	18	18	19	19	19
135	13	14	15	15	15	15	15	16

[a]Values derived from NCRP (1976) for ^{60}Co and 6 MV, and from Abrath *et al*. (1983) for 18 MV. Extrapolation to 24 MV was accomplished by comparison to primary *TVL*s.

TABLE B.5b—*TVL_1 and TVL_2 in lead (centimeters) for patient-scattered radiation at various scatter angles (based on Nogueira and Biggs, 2002).*[a]

Scatter Angle (degrees)	4 MV		6 MV		10 MV	
	TVL_1 (cm)	TVL_2 (cm)	TVL_1 (cm)	TVL_2 (cm)	TVL_1 (cm)	TVL_2 (cm)
30	3.3	3.7	3.8	4.4	4.3	4.5
45	2.4	3.1	2.8	3.4	3.1	3.6
60	1.8	2.5	1.9	2.6	2.1	2.7
75	1.3	1.9	1.4	1.9	1.5	1.9
90	0.9	1.3	1.0	1.5	1.2	1.6
105	0.7	1.2	0.7	1.2	0.95	1.4
120	0.5	0.8	0.5	0.8	0.8	1.4

[a]Biggs, P.J. (2005). Personal communication (Massachusetts General Hospital, Boston). Update of values in Nogueira and Biggs (2002).

TABLE B.6—*Mean energy (million electron volts) of patient-scattered radiation as a function of scatter angle and endpoint energy (adapted by McGinley, 2002 from Taylor et al., 1999).*

Endpoint Energy (MV)	Scatter Angle (degrees)							
	0	10	20	30	40	50	70	90
6	1.6	1.4	1.2	0.9	0.7	0.5	0.4	0.2
10	2.7	2.0	1.3	1.0	0.7	0.5	0.4	0.2
18	5.0	3.2	2.1	1.3	0.9	0.6	0.4	0.3
24	5.6	3.9	2.7	1.7	1.1	0.8	0.5	0.3

TABLE B.7—*TVLs for leakage radiation in ordinary concrete (suggested values in centimeters).*[a]

Endpoint Energy (MV)[b]	TVL_1 (cm)	TVL_e (cm)
4	33	28
6	34	29
10	35	31
15	36	33
18	36	34
20	36	34
25	37	35
30	37	36
Co-60	21	21

[a]Data for TVL_1 and TVL_2 are based on a conservatively safe adaptation of the 90 degrees (80 to 100 degrees) values of Nelson and LaRiviere (1984) and graphical extrapolations to 4 MV and 30 MV. NCRP Report No. 49 (NCRP, 1976) values used for ^{60}Co.

[b]Endpoint energy based on values from Cohen (1972).

TABLE B.8a—*Differential dose albedo (wall-reflection coefficient). Multiply each table entry by 10^{-3} (e.g., the entry 3.4 means 3.4×10^{-3}). Normal incidence on ordinary concrete, for bremsstrahlung and monoenergetic photons.*[a]

0 Degree Incidence	Angle of Reflection or Scatter (degrees) from Concrete (measured from the normal)				
Source	0	30	45	60	75
30 MV	3.0	2.7	2.6	2.2	1.5
24 MV	3.2	3.2	2.8	2.3	1.5
18 MV	3.4	3.4	3.0	2.5	1.6
10 MV	4.3	4.1	3.8	3.1	2.1
6 MV	5.3	5.2	4.7	4.0	2.7
4 MV	6.7	6.4	5.8	4.9	3.1
Co-60	7.0	6.5	6.0	5.5	3.8
0.5 MeV	19.0	17.0	15.0	13.0	8.0
0.25 MeV	32.0	28.0	25.0	22.0	13.0

[a]Table values are based on evaluation of the data from the following sources: Figures 49 and 50b in IAEA (1979), Lo (1992), and Figure 4.14(b) in NCRP (2003). The available data in the references noted were put on a common graph and conservatively safe values were selected. However, there are large uncertainties (on the order of ±50 %) in albedo values due to both the calculations and the interpolations.

TABLE B.8b—*Differential dose albedo (wall reflection coefficient). Multiply each table entry by 10^{-3} (e.g., the entry 4.8 means 4.8×10^{-3}). 45 degree angle of incidence, ordinary concrete, for bremsstrahlung and monoenergetic photons.*[a]

45 Degree Incidence	Angle of Reflection or Scatter (degrees) from Concrete (measured from the normal)				
Source	0	30	45	60	75
30 MV	4.8	5.0	4.9	4.0	3.0
24 MV	3.7	3.9	3.9	3.7	3.4
18 MV	4.5	4.6	4.6	4.3	4.0
10 MV	5.1	5.7	5.8	6.0	6.0
6 MV	6.4	7.1	7.3	7.7	8.0
4 MV	7.6	8.5	9.0	9.2	9.5
Co-60	9.0	10.2	11.0	11.5	12.0
0.5 MeV	22.0	22.5	22.0	20.0	18.0
0.25 MeV	36.0	34.5	31.0	25.0	18.0

[a]Table values are based on evaluation of the data from the following sources: Figures 49 and 50b in IAEA (1979) and Figure 4.14(b) in NCRP (2003). The available data in the references noted were put on a common graph and conservatively safe values were selected. However, there are large uncertainties (on the order of ±50 %) in albedo values due to both the calculations and the interpolations.

Table B.8c—*Differential dose albedo (wall-reflection coefficient). Multiply each table entry by 10^{-3} (e.g., the entry 5.5 means 5.5×10^{-3}). Normal incidence on iron, for bremsstrahlung.*[a]

0 Degree Incidence	Angle of Reflection or Scatter (degrees) from Iron (measured from the normal)				
Source	0	30	45	60	75
30 MV	5.5	4.7	4.4	3.8	2.3
18 MV	5.1	4.5	4.3	3.8	2.4
10 MV	5.0	4.5	4.3	3.9	2.5
6 MV	5.5	4.9	4.7	4.2	2.8
4 MV	6.0	5.4	5.1	4.8	3.1

[a]Table values are based on evaluation of the data from the following sources: Figure 50c in IAEA (1979), Lo (1992), and Figure 4.14(c) in NCRP (2003). The available data in the references noted were put on a common graph and conservatively safe values were selected. However, there are large uncertainties (on the order of ±50 %) in albedo values due to both the calculations and the interpolations.

TABLE B.8d—*Differential dose albedo (wall-reflection coefficient). Multiply each table entry by 10^{-3} (e.g., the entry 6.6 means 6.6×10^{-3}). 45 degree incidence on iron, for bremsstrahlung.*[a]

45 Degree Incidence	Angle of Reflection or Scatter (degrees) from Iron (measured from the normal)				
Source	0	30	45	60	75
30 MV	6.6	6.5	6.3	5.5	4.6
18 MV	6.5	6.4	6.2	6.0	5.6
10 MV	6.1	6.8	7.1	7.2	7.2
6 MV	6.0	7.0	8.5	9.0	9.5
4 MV	7.1	8.1	10.0	10.6	11.5

[a]Table values are based on evaluation of the data from the following sources: Figure 50c in IAEA (1979) and Figure 4.14(c) in NCRP (2003). The available data in the references noted were put on a common graph and conservatively safe values were selected. However, there are large uncertainties (on the order of ±50 %) in albedo values due to both the calculations and the interpolations.

TABLE B.8e—*Differential dose albedo (wall-reflection coefficient). Multiply each table entry by 10^{-3} (e.g., the entry 3.5 means 3.5×10^{-3}). Normal incidence on lead for bremsstrahlung.*[a]

0 Degree Incidence	Angle of Reflection or Scatter (degrees) from Lead (measured from the normal)				
Source	0	30	45	60	75
30 MV	3.5	3.0	2.7	2.4	1.5
18 MV	3.9	3.4	3.2	2.8	1.8
10 MV	4.5	3.9	3.6	3.2	2.2
6 MV	5.0	4.5	4.2	3.8	2.6
4 MV	5.9	5.2	4.7	4.2	3.0

[a]Table values are based on evaluation of the data from the following sources: Figure 50d in IAEA (1979), Lo (1992), and Figure 4.14(d) in NCRP (2003). The available data in the references noted were put on a common graph and conservatively safe values were selected. However, there are large uncertainties (on the order of ±50 %) in albedo values due to both the calculations and the interpolations.

TABLE B.8f—*Differential dose albedo (wall-reflection coefficient). Multiply each table entry by 10^{-3} (e.g., the entry 4.1 means 4.1×10^{-3}). 45 degree incidence on lead, for bremsstrahlung.*[a]

45 Degree Incidence	Angle of Reflection or Scatter (degrees) from Lead (measured from the normal)				
Source	0	30	45	60	75
30 MV	4.1	4.2	4.1	3.7	3.2
18 MV	4.9	5.0	5.0	4.8	4.5
10 MV	5.4	5.8	6.0	5.9	5.8
6 MV	6.5	6.8	7.0	7.3	7.8
4 MV	6.5	7.6	8.3	8.6	9.0

[a]Table values are based on evaluation of the data from the following sources: Figure 50d in IAEA (1979) and Figure 4.14(d) in NCRP (2003). The available data in the references noted were put on a common graph and conservatively safe values were selected. However, there are large uncertainties (on the order of ±50 %) in albedo values due to both the calculations and the interpolations.

TABLE B.9—*Neutron dose equivalent (H_0) at 1.41 m from the target per unit absorbed dose of x rays at the isocenter (mSv Gy^{-1}) and total neutron source strength (Q_n) emitted from accelerator head. A graph of Q_n as a function of nominal endpoint energy for the data in Table B.9 is presented in Figure B.1.*

Vendor	Model	Endpoint Energy (MV)		H_0	Q_n	Reference
		Nominal	Using AAPM (1983)	mSv / Gy	Neutrons per gray ($\times 10^{12}$)	
Varian	1800	18	16.6	1.02 – 1.6	1.22	McGinley (2002)
	1800	15		0.79 – 1.3	0.76	McGinley (2002)
	1800	10		0.04	0.06	McGinley (2002)
	2100C[a]	18			0.96	Followill *et al.* (2003)
	2100C[a]	18			0.87	Followill *et al.* (2003)
	2300CD	18			0.95	Followill *et al.* (2003)
	2500	24			0.77	Followill *et al.* (2003)
Siemens	KD	20	16.5	1.1 – 1.24	0.92	McGinley (2002)
	MD[a]	15		0.17	—	McGinley (2002)
	MD2	10			0.08	Followill *et al.* (2003)
	MD[a]	15			0.2	Followill *et al.* (2003)
	KD	18			0.88	Followill *et al.* (2003)

	Primus	10			0.02	Followill *et al.* (2003)
	Primus[a]	15			0.12	Followill *et al.* (2003)
	Primus[a]	15			0.21	Followill *et al.* (2003)
Philips/Electa	SL25[a]	25	22	2.0	2.37	McGinley (2002)
	SL20	20	17	0.44	0.69	McGinley (2002)
	SL20	18			0.46	Followill *et al.* (2003)
	SL25	18			0.46	Followill *et al.* (2003)
	SL25[a]	25			1.44	Followill *et al.* (2003)
GE	Saturne41	12	11.2	0.09	0.24	McGinley (2002)
	Saturne41	15	12.5	0.32	0.47	McGinley (2002)
	Saturne43[a]	18	14.0	0.55	1.50	McGinley (2002)
	Saturne43[a]	18			1.32	Followill *et al.* (2003)
	Saturne43	25	18.5	1.38	2.4	McGinley (2002)

[a]Two separate units of the same model and nominal endpoint energy.

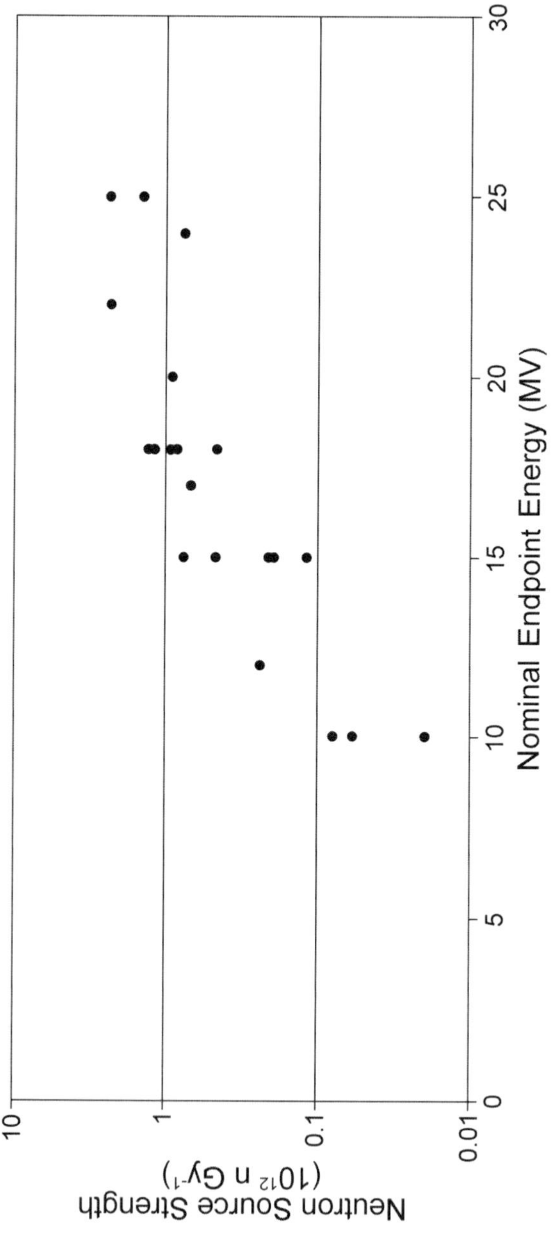

Fig. B.1. Graph of neutron source strength (Q_n) (neutrons per gray of x-ray absorbed dose at isocenter) as a function of nominal endpoint energy for data presented in Table B.9.

Appendix C

Neutron Monitoring for Radiotherapy Facilities[20]

C.1 Introduction

Neutron monitoring inside the treatment room may be performed to determine neutron leakage from the accelerator head for shielding purposes, and to determine neutron dose equivalent in the patient plane, both inside and outside the primary photon beam.

Regulatory agencies typically require shielding integrity radiation surveys during commissioning. While barriers composed of hydrogenous materials (such as concrete or earth) that provide adequate shielding for photons also provide adequate shielding for neutrons, it would be prudent to perform spot checks outside these barriers. Laminated barriers shall be monitored for neutrons beyond the shielding. Facilities operating above energies of 10 MV shall be checked for neutrons at the door, maze entrance, and any other openings through the shielding.

In neutron monitoring for radiotherapy facilities, the quantities of interest are neutron fluence,[21] neutron dose equivalent[22] (usually ambient dose equivalent),[23] or dose-equivalent rate, and the neutron spectrum (ICRU, 1993). Neutrons are classified as:

[20]Any mention of commercial products within NCRP publications is for information only; it does not imply recommendation or endorsement by NCRP.

[21]The fluence, $\Phi = \mathrm{d}N/\mathrm{d}a$, is the number of particles dN incident on a sphere of cross-sectional area da. The unit is m^{-2} or cm^{-2}.

[22]The dose equivalent H is the product of Q and D at a point in tissue, where D is the absorbed dose and Q is the quality factor at that point. Thus $H = Q\,D$. H is measured in sievert or rem.

Thermal: $\bar{E}_n = 0.025$ eV at 20 °C; typically $E_n \leq 0.5$ eV (cadmium resonance)
Intermediate: 0.5 eV $< E_n \leq 10$ keV
Fast: $E_n > 10$ keV

where E_n is the neutron energy and $\bar{E}_n$ is the average neutron energy. For therapy accelerators, it is sufficient to divide the neutron energy spectrum into only two energy regions: thermal (0 to 0.5 eV) and epithermal (>0.5 eV) (NCRP, 1984).

Neutron measurements inside the treatment room of a radiotherapy facility are fraught with difficulties because of photon interference from the primary and leakage photon beam and the fact that neutron detection is spread over many decades of energy ranging from thermal energies (0.025 eV) to several million electron volts. No single detector can accurately measure neutron fluence or dose equivalent over the entire energy range.

Most therapy linear accelerators are operated at repetition rates varying from 100 to 400 pulses per second with pulse widths of ~1 to 10 μs (AAPM, 1986b). Inside the primary beam, the photon fluence is at least 1,000 to 4,000 times higher than the neutron fluence, while outside the primary beam the photon leakage fluence is at least 10 to 100 times higher. Peak electron currents may range from 20 to 120 mA per pulse in the photon mode and 3 to 15 mA per pulse in the electron mode. The intense photon pulse usually overwhelms any active detector that detects particle events electronically. Thus the measured readings are only an indication of the repetition rate of the accelerator. For moderated detectors, it has been observed that the measured readings are higher than the repetition rate due to scattered radiation within the room, and the fact that the neutron moderation time allows an event to be detected after the pulse has ended.[24] It may be possible to adjust the accelerator to a low output per pulse, but there is no guarantee that

[23]The ambient dose equivalent $H^*(d)$, at a point in a radiation field, is the dose equivalent that would be produced by the corresponding expanded and aligned field in the ICRU sphere at a depth d, on the radius opposing the direction of the aligned field. The ambient dose equivalent is measured in sievert or rem. For strongly penetrating radiation, a depth of 10 mm is recommended. In the expanded and aligned field, the fluence and its energy distribution have the same values throughout the volume of interest as in the actual field at the point of reference, but the fluence is unidirectional.

[24]Thomas, D.J. (2005). Personal communication (National Physical Laboratory, Teddington, United Kingdom).

the photon to neutron ratio remains the same. Another problem encountered in mixed photon-neutron fields is that the neutron detectors can have photon-induced reactions, which cannot be separated from the neutron interactions themselves. This interference should be considered when measuring neutrons in the primary photon beam. Thus only passive detectors should be used for measurements inside the room, except perhaps at the outer maze entrance area.

Outside the shielded treatment room, both the neutron and photon fluences are considerably lower. Further, the neutron pulse is spread over several hundred microseconds because of moderation in the shielding material. Therefore, active detectors may be used with the caveat that photon pulse pileup and dead time effects on their responses should be considered. Passive detectors with high sensitivities can also be used.

Many of the neutron detectors that measure fluence require the use of fluence to dose-equivalent conversion coefficients to obtain the dose equivalent. These coefficients depend strongly on the neutron energy and hence the neutron spectrum must be known or approximated (Nath *et al.*, 1979). The photoneutrons emitted from the accelerator head of a therapy linear accelerator consist of direct emission neutrons and evaporation neutrons (NBS, 1964). Direct emission neutrons have average energies of a few million electron volts. The angular distribution of these neutrons follows a $\sin^2 \theta$ distribution. For heavier nuclei this process contributes to ~10 to 20 % of the photoneutron yield for bremsstrahlung spectra with upper energies ranging from 15 to 30 MeV. Most neutrons are produced as a result of the evaporation process or giant photonuclear resonance in heavy nuclei (atomic mass number A greater than ~40) (IAEA, 1979). Evaporation neutrons are emitted isotropically. The spectral distribution of these neutrons is nearly independent of the photon energy absorbed for photons more than a few million electron volts above the neutron production threshold. In general, the average energy generally lies in the range of 1 to 2 MeV.

McCall and Swanson (1979) have shown that the integral photoneutron spectra for 15 MeV electrons on a tungsten target is quite similar to the ^{252}Cf fission neutron spectrum, with the average energy of the ^{252}Cf neutrons being slightly higher. The photoneutron spectrum from the accelerator head changes after penetration through the head shielding (NCRP, 1984). Since the linear accelerator is usually in a concrete shielded room, room-scattered neutrons will further soften the spectrum. The photoneutron spectrum outside the concrete room resembles that of a heavily shielded fission spectrum, and the average energy is significantly lower than that inside the room. Most of the neutrons are <0.5 MeV in energy. The average energy of neutrons at the outer maze entrance is roughly ~100 keV.

Many of the neutron detectors are normally calibrated against a plutonium-beryllium ($\bar{E}_n$ = 4.2 MeV), americium-beryllium ($\bar{E}_n$ = 4.5 MeV), or ^{252}Cf ($\bar{E}_n$ = 2.2 MeV) neutron source. Since plutonium-beryllium and americium-beryllium neutron sources have higher average energies, their use can lead to systematic uncertainties. The spectrum of fission neutrons from ^{252}Cf is in general similar to a typical photoneutron spectrum, and the shape of the photoneutron spectrum is rather independent of the incident electron energy. Thus a detector calibrated against a ^{252}Cf source may be adequate for neutron measurements inside the primary beam. However the spectrum outside the primary beam and outside the room shielding is a heavily shielded photoneutron spectrum, so the assumption of a fission spectrum may lead to significant errors (McCall *et al.*, 1978).

Various active and passive neutron monitoring detectors and methods are discussed in the subsequent sections, including some detectors that are simple to use and commercially available in the United States. The response characteristics of neutron survey instruments available in the United Kingdom have been reported elsewhere (Bartlett *et al.*, 2002).

C.2 Neutron Monitoring Techniques

Neutron monitoring techniques for radiotherapy facilities consist of active and passive methods. Active methods include the use of rem- and fluence meters. Passive methods include the use of activation foils, thermoluminescent dosimeters (TLDs), solid-state nuclear track detectors, and bubble detectors.

C.2.1 *Active Monitoring*

Active neutron monitoring usually relies on slowing down fast neutrons or moderating them until they reach thermal energies. A thermal detector is then used to detect the thermal neutrons. Depending on the geometry and configuration of the moderator, the instrument is designed to measure either dose equivalent (commonly called rem-meters) or fluence (fluence meters).

C.2.1.1 *Rem-Meters.* Active detectors such as neutron rem-meters are useful in radiation fields for which the neutron spectrum is not well characterized, since their response is designed to be proportional to the dose equivalent and therefore independent of neutron energy. Thus, in principle, no knowledge of the neutron spectrum is required. Although rem-meters may vary significantly in size and geometrical configuration the underlying principle of operation is similar (Olsher *et al.*, 2000). The response of a rem-meter is shaped to fit an appropriate fluence to dose-equivalent conversion coefficient (referred to as "factor" in the past) over a particular energy range. In the older designs, ICRP Publication 21 (ICRP, 1973) fluence to dose-equivalent conversion coefficients were used. These conversion coefficients were also recommended by NCRP Report No. 38 (NCRP, 1971), and are the basis for the federal and various state regulations in the United States. The newer designs of rem-meters comply with ICRP effective dose recommendations (ICRP, 1991). The operational quantity appropriate for rem-meter calibration is ambient dose equivalent [$H^*(10)$] which is defined in Equation C.1 for a known neutron spectrum:

$$H^*(10) = \int h_{\Phi}(E)\ \Phi(E)\ \mathrm{d}E \qquad \text{(C.1)}$$

In Equation C.1, $h_{\Phi}(E)$ is the fluence to ambient-dose-equivalent conversion function, and $\Phi(E)$ is the neutron fluence as a function of energy for a given neutron field. The rem-meter response (R_m) in that field is given by Equation C.2:

$$R_m = \int \mathrm{C}\ r_{\Phi}(E)\ \Phi(E)\ \mathrm{d}E \qquad \text{(C.2)}$$

In Equation C.2, $r_{\Phi}(E)$ is the response function of the rem-meter in counts per unit fluence, and C is the calibration constant in units of sievert per count. The rem-meter measurement is considered accurate as long as $r_{\Phi}(E)$ has a similar energy response to that of $h_{\Phi}(E)$. The ratio $r_{\Phi}(E) / h_{\Phi}(E)$ defines the traditional energy response of the rem-meter in terms of counts per unit dose equivalent. The main problem is that $r_{\Phi}(E)$ does not match $h_{\Phi}(E)$ over much of the energy range. Therefore, some detectors tend to either over-respond or under-respond in certain energy regions. Typically, most rem-meters have a very large over-response in the intermediate energy region, and give an adequate measure of dose equivalent between 100 keV and 6 MeV (Rogers, 1979). Therefore, it is important to know the spectrum, at least roughly, before any reliance can be placed on the instrument readings.

Most commercial rem-meters consist of a neutron moderator (usually hydrogenous material such as polyethylene) surrounding a thermal neutron detector [*e.g.*, a ^{10}B enriched BF_3 detector or ^3He counter tube or ^6LiI (europium) scintillator] (NCRP, 2003). The moderator slows down fast and intermediate energy neutrons, which are then detected by the thermal neutron detector. The reactions that take place in each of these detectors are as follows:

BF_3 proportional counter: $^{10}B(n_{th},\alpha)^7Li$, E_Q = 2.31 MeV, σ = 3,840 barns
^3He proportional counter: $^3He(n_{th},p)^3H$, E_Q = 0.765 MeV, σ = 5,330 barns
$^6LiI(Eu)$ scintillator: $^6Li(n_{th},\alpha)^3H$, E_Q = 4.78 MeV, σ = 940 barns

where E_Q is the kinetic energy released and σ is the cross section for the thermal neutron reaction. The cross sections drop roughly as $E_n^{-1/2}$. Thus at 1 MeV the cross section is about four orders of magnitude lower than at thermal energies. When used without a moderator these detectors are sensitive only to thermal neutrons. Because of their excellent photon rejection and low cost, rem-meters incorporating BF_3 proportional counters are more commonly used outside the shielding in therapy facilities. The ^3He proportional counter is more sensitive and more stable than the BF_3 counter, but it is much more expensive. The $^6LiI(Eu)$ scintillator has a very high sensitivity but very poor photon rejection, thus making it difficult to use in mixed photon-neutron fields such as those encountered at therapy facilities. It is important to note that for all active monitors, manufacturers normally state photon rejection for steady fields and not pulsed fields.

In the BF_3 detector, the thermal neutrons are captured in the boron *via* the $^{10}B(n_{th},\alpha)^7Li$ reaction. The alpha particle and recoil ^7Li nucleus each produces a large pulse in the proportional counter. The large pulses are orders of magnitude higher than the pulses produced by photon interactions, and therefore can be discriminated from the small pulses produced by photons in mixed fields.

The energy response of a moderated detector is determined mainly by its size and geometrical configuration (NCRP, 2003). Rem-meter designs have evolved over the years. One of the more commonly used rem-meters is the Andersson-Braun rem-meter (AB-meter), which has a cylindrical polyethylene moderator surrounding a BF_3 counter tube (Andersson and Braun, 1963). This arrangement results in a dose-equivalent-like response from thermal to ~10 MeV, except at intermediate energies where the rem-meter over-responds. A borated plastic sleeve is used around the counter tube to minimize the over-response in the 10 to 100 keV range. Holes drilled in the sleeve increase the response at thermal energies. However, the cylindrical moderator geometry of the instrument impacts the directional response. A change in response with instrument orientation of as much as 35 % for a neutron energy of 1 MeV has been observed (Cosack and Lesiecki, 1985). A 65 % underestimate in the true dose equivalent has been observed when the rem-meter is exposed with its side oriented to a source of thermal neutrons (Hankins and Cortez, 1974). In order to improve the directional dependence, a modified version with a spherical polyethylene moderator and a cadmium layer was designed by Hankins and is marketed by Thermo Electron Corporation (Sante Fe, New Mexico) as the model NRD (Section C.2.1.2). The high-energy response of this detector decreases steadily above 7 MeV, but at these energies, the neutron fluence is considerably reduced for medical accelerators.

C.2.1.2 *Commercial Instruments*. Figure C.1 shows the Victoreen neutron survey meter Model 190N (Fluke Biomedical Radiation Management Services, Cleveland Ohio), which is based upon the classical Andersson-Braun rem-meter design. The moderator consists of a polyethylene cylinder 24 cm (9.5 inches) long and 21.6 cm (8.5 inches) in diameter surrounding the BF_3 tube. The fill gas is 96 % enriched ^{10}B. The instrument is available with display in either rem or sievert. According to the manufacturer, the instrument has a dose-equivalent range from 0 μSv h^{-1} to 0.75 Sv h^{-1}, and it can integrate from 0 μSv to 10 Sv. It has a gamma rejection up to ~5 Gy h^{-1} for ^{137}Cs and the isotropy is <20 % in orthogonal directions. The rem-meter weighs 9.52 kg (21 pounds).

Figure C.2 shows a Thermo Electron Corporation ASP/2e NRD neutron survey meter[25] that is operated with a portable battery. The detector is a 22.9 cm (9 inch) diameter, cadmium-loaded polyethylene sphere with a BF_3 tube in the center. According to the manufacturer, this detector has been shown to have a response that closely follows theoretical dose equivalent from neutrons over the energy range from 0.0025 eV (thermal) to ~10 MeV. The dose-equivalent range is 1 to 100 mSv h^{-1}. The BF_3 tube provides gamma rejection up to ~5 Gy h^{-1}, depending on the voltage that is used (1,600 to 2,000 V). The instrument has a dead time of 10 μs (nominal). It has an isotropic response within 10 %. The response time is programmable from 0 to 255 ms. The accompanying counting instrument is the Model ASP/2e, which has a dual

[25]Previously named the Thermo Eberline ASP/2e NRD.

Fig. C.1. Victoreen neutron survey meter Model 190N.

Fig. C.2. Thermo Electron Corporation ASP/2e NRD neutron survey meter.

analog/digital display. It has a rate meter that can be used in an integrate or scaler mode. It has a count range of 1 to 1.3 million counts per minute.

Figures C.3 and C.4[26] show the response functions for the Andersson-Braun rem-meter and the NRD neutron survey meter. Figure C.3 is based on NCRP Report No. 38 fluence to dose-equivalent conversion coefficients (NCRP, 1971). The maximum

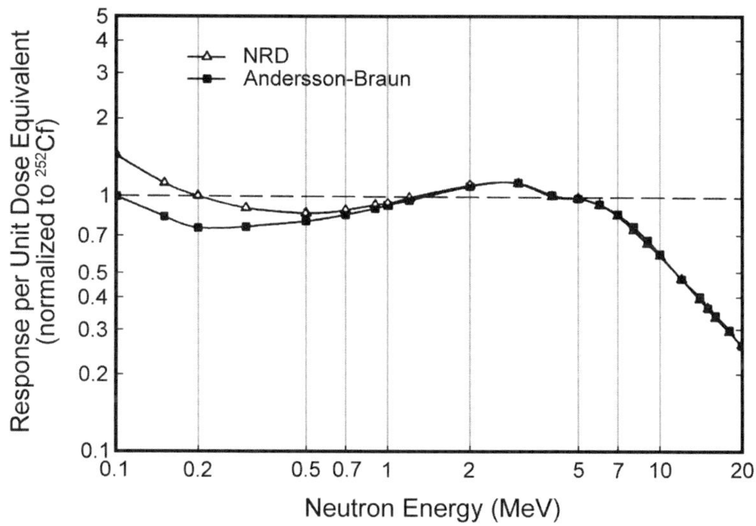

Fig. C.3. Response per unit dose equivalent of Thermo Electron Corporation NRD neutron survey meter and Andersson-Braun rem-meter.[26]

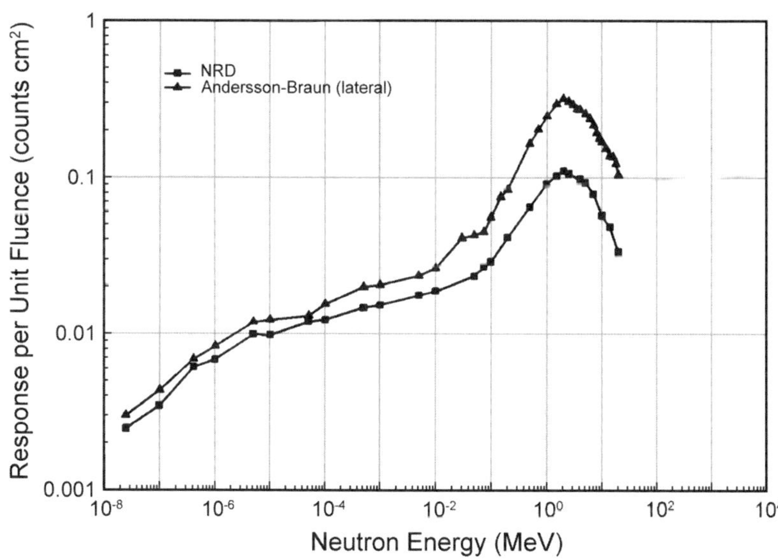

Fig. C.4. Response per unit fluence of Thermo Electron Corporation NRD neutron survey meter and Andersson-Braun rem-meter.[26]

[26]Olsher R.H. (2005). Personal communication (Los Alamos National Laboratory, Los Alamos, New Mexico).

neutron energy for a 15 MV photon beam is ~10 MeV. Since the response falls off rapidly at energies above 6 MeV, these detectors are not necessarily accurate over the energy range of interest in radiotherapy facilities, but are adequate for monitoring purposes. Pulse pileup at high photon dose-equivalent rates and dead time corrections at high neutron dose-equivalent rates need to be considered.

For pulsed radiation with a pulse repletion rate of P_r pulses per second, and if the dead time T_D is shorter than the pulse width T_P, the corrected number of counts (C_{corr}) is related to the measured number of counts (C_n) by the following expression (IAEA, 1979):

$$C_{corr} = \frac{C_n}{1 - \frac{C_n\, T_D}{P_r\, T_P}\left(1 - \frac{T_D}{2\, T_P}\right)} \tag{C.3}$$

If T_D is longer than T_P:

$$C_{corr} = C_n - P_r \ln\left(1 - \frac{C_n}{P_r}\right) \tag{C.4}$$

For a moderated thermal-neutron detector, the effective dead time is determined by the moderation time, and not the pulse width for the neutrons. Therefore, the pulse width is replaced by the moderation time in the above equations. For spherical moderators with diameters of 30 cm and 45 cm, the neutron moderation time is of the order of 200 μs. For a distribution of neutron moderation in different types of rem counters, the reader is referred to IAEA (1979). For well-shielded facilities, the neutron dose-equivalent rates are fairly low and may be of the order of 1 $\mu Sv\ h^{-1}$. Therefore, integrated measurements of dose equivalent are preferable to instantaneous dose-equivalent rate measurements. As with any radiation detector, a correct response is obtained only when the radiation field irradiates the entire detector (*i.e.*, the size of the radiation field is larger than the size of the detector). If the size of the radiation field is smaller than the size of the detector, a reduced response will be obtained, as is the case of measurements through gaps or small holes in the shield. This may require the delivery of a large amount of monitor units at the isocenter.

C.2.1.3 *Fluence Meters.* The instrument most generally used for neutron-fluence measurement is a version of the BF_3 long counter (Hanson and McKibben, 1947). Its principle of operation was discussed in Section C.2.1.2. The bare BF_3 counter is sensitive to thermal neutrons only. The sensitivity of a counter is proportional to volume, pressure and degree of enrichment in ^{10}B. When used with a moderator, the fast neutrons are thermalized and detected. If the moderator is enclosed in 0.5 mm of cadmium, the thermal neutrons are completely eliminated. The thickness of hydrogenous moderator can be chosen such that a fairly flat response is obtained per unit fluence for neutron energies up to several million electron volts. The moderator is usually cylindrical. Once the neutron fluence is known, an appropriate fluence-to-dose equivalent conversion coefficient can be applied, based on an average energy, to obtain dose equivalent (NCRP, 1984). Thus the use of the BF_3 detector requires knowledge of the neutron spectrum. It is particularly handy when used in conjunction with rem-meters when neutron dose-equivalent rates are below measurable levels on the rem-meter. The moderated BF_3 detector is useful to monitor relative variations of the neutron field with time, and is especially useful for measurements during intensity-modulated radiation therapy. Further, the ratio of the rem-meter and moderated BF_3 detector readings can be used to obtain a rough estimate of the average energy of the neutron spectrum. All the problems associated with rem-meters also apply to these detectors.

C.2.1.4 *Neutron Spectrometers.* Thermal-neutron detectors can be used inside a series of hydrogenous spheres of varying diameters to determine the neutron spectrum (Bramblett *et al.*, 1960). Since the amount of moderation varies in each of these spheres, it is possible to calculate the total spectrum by taking all the responses and folding them into a series of equations. Since this usually requires a computer program, a large number of spheres and long measurement times, the process is fairly laborious. This method is referred to as the Multi-Sphere or Bonner-Sphere method.

An alternate approach is to use a scintillation spectrometer with either a plastic or liquid scintillator (McCall, 1981). Fast neutrons can be detected by the scintillation light from proton recoil in hydrogenous scintillators (Knoll, 1989). The energy distribution of the recoil protons varies from zero to the full neutron energy and their range is usually small compared to the dimensions of the scintillator. Thus their full energy is deposited in the scintillator and the expected pulse-height distribution is rectangular. However, as discussed before, in a mixed photon-neutron field, the interference from photons can be significant, and requires a considerable amount of time in adjusting the pulse-shape discrimination to reject the photon signal. The difficulties associated with the procedure preclude its use for routine neutron measurements around radiotherapy facilities. Further, it is also affected by the pulsed nature of the neutron field.

Time-of-flight spectrometers normally provide the best measurements of neutron spectra. A signal is produced at the point of creation of the neutron or when it first enters the detection system, and the time is measured until that neutron gets to a detector some distance away. The energy of the neutron can be determined by knowing the time taken to travel a given distance. The time-of-flight method may not be applicable at radiotherapy facilities because of the scattering of the majority of neutrons after they are created. Further, it is not clear that a suitable "start signal" may be obtained. These detectors are also susceptible to pulsed-field effects. Additionally, this is a difficult measurement requiring expensive detectors, and therefore not very practical for neutron monitoring at radiotherapy facilities.

C.2.2 *Passive Monitoring*

Passive monitoring is the method of choice for neutron measurements inside the treatment room. Passive monitors with high neutron sensitivity can also be used outside the room. The various types of passive monitors are: moderated thermal-neutron activation detectors, threshold activation detectors, TLDs, solid-state nuclear track detectors, and bubble detectors. Passive thermal-neutron monitors such as activation foils and TLDs can also be used inside a series of hydrogenous spheres of varying diameters to determine the neutron spectrum, as discussed in Section C.2.1.4. Such systems have been described in the literature (Axton and Bardell, 1979; Nath *et al.*, 1979; Thomas *et al.*, 2002). When using passive detectors, it is important to take several measurements with the same detector type at a given location to achieve adequate statistics and error analysis.

C.2.2.1 *Activation Detectors.* Activation detectors include thermal-neutron detectors and threshold detectors. A bare foil and a cadmium-covered foil can be used to determine thermal neutron fluences. Activation foils are completely stable and reproducible (Axton and Bardell, 1979).

Moderated thermal-neutron detectors are used to determine fast-neutron fluence. The moderator provides an energy independent thermal neutron fluence that is proportional to the incident fast-neutron fluence. Neutron absorption by the foil results in the production of a radioactive nucleus. Measurement of the radioactivity can be correlated with the thermal-neutron fluence incident on the activation foil at the center of the moderator. Activation foils such as gold and indium have a high cross section for thermal-neutron absorption as shown in Table C.1 (AAPM, 1986b). Fast-neutron interactions that result in radioactivity have much smaller cross sections (usually <1 barn) than the thermal-neutron capture processes. Detectors based on this principle are called threshold detectors because there is a threshold above which the reaction takes place. A bare activation detector of phosphorous may be used to measure both thermal and fast neutron fluences utilizing the reactions shown in Table C.1. The thermal neutron reaction in phosphorous has a very low cross section compared to gold and indium. Details of the measurement techniques can be found in AAPM (1986b). The extent of photon interference on these activation detectors must also be considered.

The resonance peaks for thermal-neutron capture in indium and gold are similar in energy and magnitude. They peak at a few electron volts above the cadmium cut-off energy (0.4 eV). Because of its large cross section and short half-life ($T_{1/2}$), indium is much more sensitive than gold. It can be reused after several hours, since it has a short half-life. It is also less expensive. However, it has to be read out fairly quickly after irradiation. Indium is very soft but it can be obtained in the preferred 0.125 mm thickness, mounted on an aluminium support. The photon-induced responses in the aluminium-mounted indium lead to ^{26}Al ($T_{1/2} = 7 \times 10^5$ y) and ^{26m}Al ($T_{1/2}$ = 6.34 s) *via* the (γ,n) reaction. But because of the small cross section (16 barns) for the reaction and the half-lives associated with the products, the interference is negligible after a few minutes of waiting time. Gold on the other hand is much less sensitive, more expensive and requires several weeks of waiting between irradiation and reuse. However, because of its long half-life, there can be a substantial time delay between irradiation and counting.

The activation foils and moderators can be purchased commercially.[27] The foils need to be thin enough to minimize self-absorption. The moderators are cylindrical, typically ~15 cm in diameter and 15 cm in height, and are covered with either a thin layer of cadmium or boron, to allow for the separation of thermal neutrons and those neutrons above the cadmium cut-off. The moderator foil system is normally placed with its center at the point to be measured. For measurements inside the primary beam, the field size should be wide enough to irradiate the entire moderator.

If several moderators are used, care should be taken to ensure that the distance between the moderators is at least twice the diameter of the moderator. For in-beam measurements, photon-induced effects in the moderator and cadmium lining need to be

[27]Thermo Electron Corporation, Sante Fe, New Mexico.

TABLE C.1— *Characteristics of activation foils suitable for use at radiotherapy facilities (AAPM, 1986b).*

Reaction	Cross Section (barns)	Percent Abundance	Product Half-Life	Decay Radiation (MeV)	Branching Intensity	Use
$^{115}In(n_{th},\gamma)^{116m}In$	194	95.7	54 m	β^-: 1.00 γ: 0.138 – 2.111	1.00	Inside room; inside and outside primary beam with moderator
$^{197}Au(n_{th},\gamma)^{198}Au$	99	100	2.698 d	β^-: 0.962 γ: 0.412	0.99 0.99	Inside room; inside and outside primary beam with moderator
$^{31}P(n,p)^{31}S$ threshold = 0.7 MeV	varies with energy	100	2.62 h	β^-: 1.48 γ: 1.26	0.99 0.07	Inside room; inside primary beam
$^{31}P(n_{th},\gamma)^{32}P$	0.190	100	14.28 d	β^-: 1.71	1.00	Inside room; inside primary beam

considered (AAPM, 1986b; McCall *et al.*, 1976). Since these corrections can be large for cadmium-covered moderators and at higher photon energies, this method is not recommended for bremsstrahlung endpoint energies above 20 MV. Boron-lined moderators are less susceptible to photon-induced effects. Gold and indium foils can be counted for beta decay with a thin-window Geiger-Mueller tube or a proportional counter. They can also be counted with a scintillation counter or germanium (lithium) detector by detecting the gamma rays. The disadvantage of moderated activation foils is their large size.

Phosphorous is used normally in the form of phosphorous pentoxide powder. All unwanted activation products from photon and neutron fields are short lived and therefore of no importance in the analysis of ^{31}S and ^{32}P. Neutrons produced in the phosphorous pentoxide are captured by the ^{31}P nuclei, however this interference is <5 %. Since the activation products from phosphorous are beta emitters, liquid-scintillation counting is the preferred method though it is not found routinely in nonacademic medical environments. For in-beam measurements, the phosphorous technique yields neutron dose equivalents that are ~98 to 99 % of the true value (AAPM, 1986b). The disadvantage of this method is that it is fairly laborious since the phosphorous pentoxide powder has to be first weighed, then carefully dissolved in distilled water in a well-ventilated area, preferably within a hood. A portion of the solution then needs to be pipetted into a liquid-scintillation vial containing liquid-scintillation fluid, and then placed in a refrigerated and automated liquid-scintillation counter. Counting is initiated only after a waiting period of ~0.5 h. A series of counts have to be taken to determine the ^{31}S activity, followed by a 12 to 24 h waiting period, after which a series of counts have to be taken to determine the ^{32}P activity.

C.2.2.2 *Solid-State Nuclear Track Detectors.* Solid-state nuclear track detectors, such as CR-39® (made by the polymerization of the monomer allyl glycol carbonate), detect neutrons mainly by sub-microscopic damage trails from the recoil nuclei of its constituent atoms, namely hydrogen, carbon and oxygen. The damage trails or tracks can be revealed by a suitable etching process (chemical etching, electrochemical etching, or both combined). On the basis of calibration exposures, the track density can then be related to the neutron dose equivalent. The use of a hydrogenous radiator in contact with the CR-39® can enhance its response because of the additional protons generated within the radiator. When combined with radiators, CR-39® can detect neutrons over a wide energy range from ~100 keV to 20 MeV. The energy range, dose-equivalent response and the lower limit of detection depend on the material, the type and thickness of the radiator, and the particular etching process used. The detectors suffer from directional dependence. CR-39® detectors should not be used for measurements inside the primary photon beam because of the following photon-induced effects (Ipe *et al.*, 2000):

^{2}D(γ,n)p, E_{th} = 2.23 MeV (E_{th} = threshold energy; 0.02 % of hydrogen is deuterium)
^{16}O(γ,α)^{12}C, E_{th} = 7.2 MeV
^{12}C(γ,α)^{8}Be $\rightarrow$ 2α, E_{th} = 7.4 MeV

One of the commercially available CR-39® detectors suitable for use at radiotherapy facilities is the Neutrak 144® (Landauer, Inc., Glenwood, Illinois), which comes with two options. The fast-neutron option uses a polyethylene radiator for fast neutrons that records recoil protons resulting from neutron interactions in the dosimeter. The thermal-neutron option has a dosimeter design intended for both fast and thermal neutrons. The left area of the chip uses a polyethylene radiator for fast neutrons while the right area uses a boron-loaded Teflon® (E.I. DuPont de Nemours and Company, Wilmington, Delaware) radiator for both fast and thermal neutrons that also records alpha particles resulting from neutron interactions in the dosimeter. The CR-39® is etched in a chemical bath to enlarge the tracks. The tracks are then counted in an automatic counter. The track density is a measure of the neutron dose equivalent. The fast-neutron dose equivalent is measured by counting the tracks generated as a result of the proton recoil with the polyethylene radiator, while the thermal-neutron dose equivalent is measured by counting the alpha tracks generated with the boron radiator. The energy range and the minimum threshold of detection (given in parentheses) are as follows: fast neutrons from 40 keV to over 35 MeV (0.2 mSv); thermal neutrons <0.5 eV (0.1 mSv).

The Neutrak ER (extended range) dual-element dosimeter combines a TLD albedo dosimeter with Neutrak 144® to allow detection of an extended range of neutron energies (*i.e.*, TLD albedo: 0.5 eV to 100 keV). It is important to note that these detectors are normally used for personal dosimetry and are therefore calibrated on a phantom, thus free-in-air exposures will result in some uncertainties. Figure C.5 shows the sensitivity of the Neutrak 144® as a function of energy.

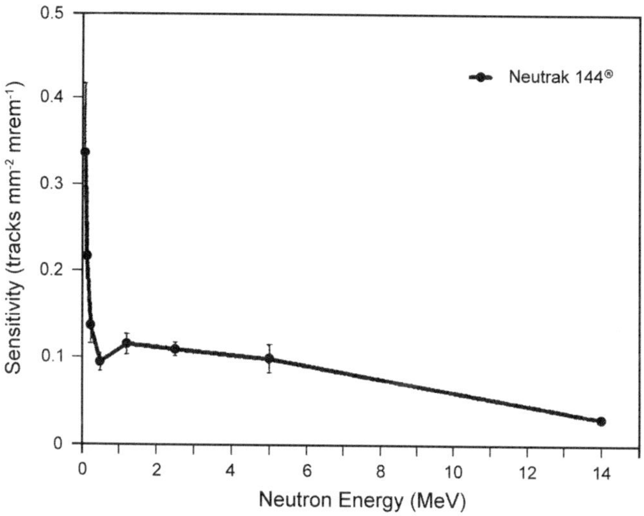

Fig. C.5. Sensitivity of Neutrak 144® as a function of neutron energy on a polymethylmethacrylate phantom (courtesy of Landauer Inc., Glenwood, Illinois). 1 mrem = 1 μSv.

C.2.2.3 *Bubble Detectors.* A bubble detector consists of tiny superheated droplets that are dispersed throughout a firm elastic polymer contained in a small sealed tube. The detector is sensitized by unscrewing the cap. Secondary charged particles are produced when the neutrons strike the droplets. The energy deposited by the charged particles causes the droplets to vaporize, producing bubbles which remain fixed in the polymer. The bubbles can be counted by eye or in an automatic reader. An inexpensive method of counting is to project the image of the bubble detector *via* a video camera on to a television screen, and then count by eye or score bubbles on a clear transparency sheet placed over the television screen (Ipe and Busick, 1987; Ipe *et al.*, 1988). The bubble detector should be rotated at least four times and counted again. Another method is to take a photograph of the detector containing the bubbles and counting them on the photograph or on an enlarged copy.[28] The number of bubbles is proportional to the neutron dose equivalent.

The advantages of the bubble detectors are that they are easy to use, have very high sensitivities, are reusable, are integrating, allow instant visible detection of neutrons, and have an isotropic response. Bubble detectors can be used for neutron monitoring

[28]Ing, H. (2005). Personal communication (Bubble Technology Industries, Chalk River, Ontario, Canada).

inside and outside the treatment room but not inside the primary photon beam (Ipe and Busick, 1987; Ipe *et al.*, 2000). The disadvantages are that sensitivities do vary within a given batch, and occasionally spurious results are obtained. However, according to the manufacturer, these problems are being resolved with improved manufacturing processes.[27] Further, there may be a loss of linearity at the higher dose equivalents, which according to the manufacturer may be attributed to a change of pressure within the gel when there are a large number of bubbles present. Finally, it is difficult to get good statistics. Therefore, it is important to use at least a minimum of three detectors at each measurement location.

Figure C.6 shows a photograph of bubble detectors available from Bubble Technology Industries (Chalk River, Ontario, Canada). The bubble detectors are normally calibrated by the manufacturer against an americium-beryllium source. The bubble detector-personal neutron dosimeter (BD-PND) is a temperature-compensated bubble detector and therefore the most useful for monitoring since no temperature corrections are required, as is the case for the BD-100R (bubble detector, threshold = 100 keV, reusable). According to the manufacturer the BD-PND has an approximate threshold of 100 keV with a reasonably flat dose-equivalent response from ~200 keV to >15 MeV. Since these detectors come in varying sensitivities, the ones with lower sensitivity can be used inside the treatment room while the most sensitive ones can be used outside the shielded facility. The energy response of the BD-PND is shown in Figure C.7. By varying the formulation, bubble detectors with different thresholds have been made. The bubble detector spectrometer (BDS) can be used for spectral measurements. Figure C.8 shows the response per unit fluence of the BDS. Measurements made by Ongaro *et al.* (2000) for two radiotherapy facilities (15 MV Siemens and 18 MV Elekta linear accelerators), indicate that BDS spectral measurements in the patient plane are within ~20 % of Monte-Carlo calculations.

The bubble detector thermal uses a ^{6}Li compound dispersed throughout the polymer and a special formulation to detect preferentially alpha particles from the $^{6}Li(n,\alpha)^{3}H$ reaction (Ing *et al.*, 1996). The thermal-to-fast neutron ratio is ~10:1. Table C.2 shows the characteristics of the various types of bubble detectors available from Bubble Technology Industries.

C.2.2.4 *Comparison of Various Passive Monitors.* Figure C.9 shows a comparison of neutron dose equivalent measured by different neutron dosimeters in the patient plane of a 15 MV Varian Clinac 2300C/D linear accelerator, for a field size of 20 cm × 20 cm. The

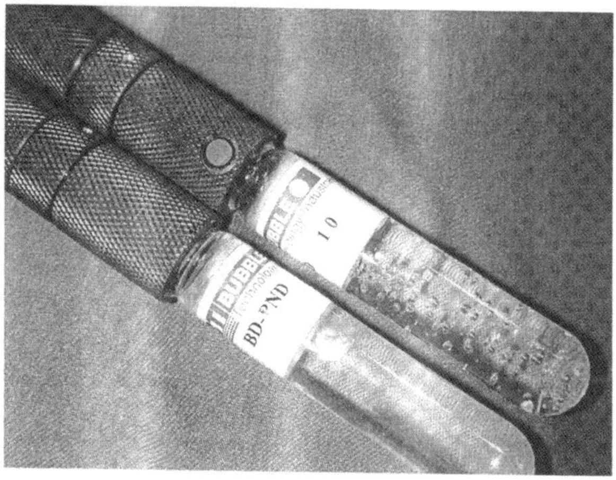

Fig. C.6. BD-PND bubble detectors (courtesy of Bubble Technology Industries, Chalk River, Ontario, Canada).

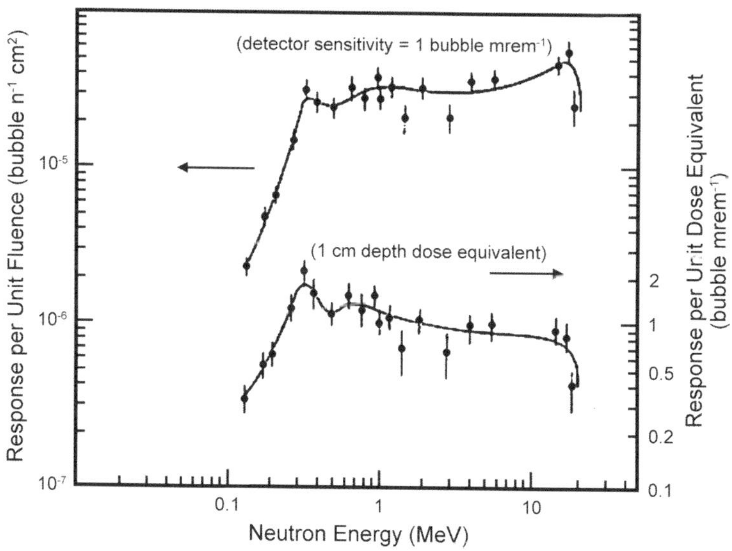

Fig. C.7. Response of BD-PND as a function of energy (Ing *et al.*, 1997). Normalized response per unit fluence (●) and per unit dose equivalent (♦). Conversion from fluence to dose equivalent based on NCRP (1971). 1 mrem = 10 μSv. "1 cm depth dose equivalent" means the dose-equivalent values at 1 cm depth, using the depth dose curves in NCRP (1971).

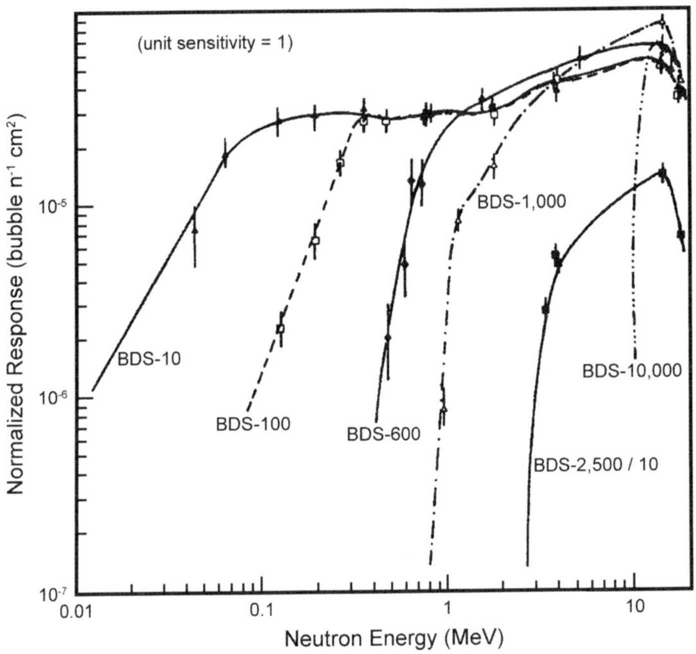

Fig. C.8. Normalized response of BDS (Ing *et al.*, 1997). The numbers that follow BDS are threshold energies in kiloelectron volts. The number 2,500/10 means that on the graph, the response of the BDS-2,500 has been reduced by a factor of 10.

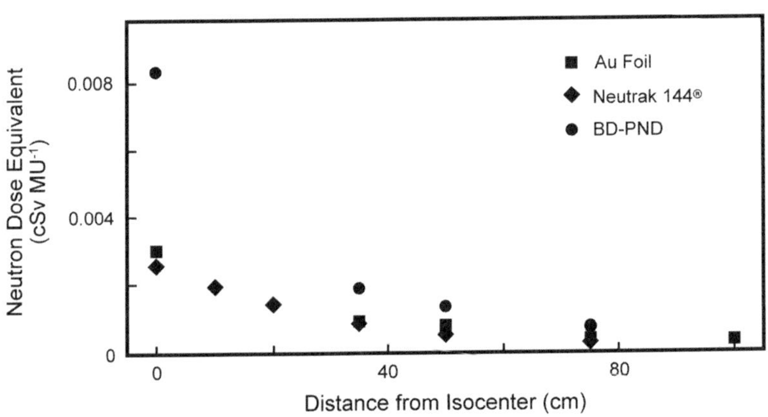

Fig. C.9. Comparison of neutron dose equivalents measured by different dosimeters. Neutron dose equivalent in patient plane for 15 MV Varian Clinac 2300 C/D (Ipe *et al*, 2000).

gold foils were 0.025 mm thick. The moderator was a cylinder of low-density polyethylene (diameter = 15.9 cm, length = 15.7 cm) covered with a boron shield which absorbs thermal neutrons. The fast-neutron fluence was converted to dose equivalent by means of appropriate conversion coefficients (NCRP, 1984). There is reasonable agreement between the moderated gold foil and the Neutrak 144® data, with slight differences being attributable to the differences in energy responses and the differences in calibration sources. Inside the primary photon beam, the BD-PND overresponds by about a factor of three. The BD-PND responses outside the primary beam are just slightly higher than the moderated gold foil and the Neutrak 144® responses. Both the BD-PND and Neutrak 144® were calibrated against an americium-beryllium source by the manufacturers. The moderated gold foils were calibrated against a ^{252}Cf source.

It is important to perform photon measurements when performing neutron measurements, since total dose equivalent is required.

C.3 Conclusions

Various neutron-monitoring techniques for radiotherapy facilities have been discussed. Due to the pulsed nature of the accelerator beam and the high-intensity photon field, active detectors should only be used outside the shielded treatment room, and possibly in the outer maze area. Passive monitoring, such as phosphorous activation and moderated gold and indium foils, can be used inside the treatment room, and inside the primary beam with the caveat that the moderated activation foils should not be used above 20 MV. Moderated activation foils can also be used inside the treatment room, outside the primary beam. Neutrak 144® and bubble detectors can be used inside the treatment room, outside the primary beam. Because of their high sensitivity, bubble detectors can also be used for radiation surveys outside the shielded treatment room.

TABLE C.2—*Characteristics of bubble detectors (BTI, 2005).*

Characteristics	BD-PND	BD-100R	BDT[a]	BDS
Energy range	<200 keV to >15 MeV	<200 keV to >15 MeV	Thermal (~1/*v* for epithermals)	6 distinct thresholds: 10; 100; 600; 1,000; 2,500; 10,000 keV
Dose-equivalent range or nominal estimate	0.01 to 50 μSv	0.01 to 50 μSv	0.01 to 1 μSv	~5 μSv
Sensitivity (user selectable)	0.033 to 3.3 bubbles per μSv	0.033 to 3.3 bubbles per μSv	3 bubbles per μSv	0.1 to 0.2 bubbles per μSv
Gamma sensitivity	None	None	None	None
Tissue equivalence	Yes	Yes	Yes	Yes
Temperature compensation	Yes	No	Yes	No
Optimum temperature range	20 to 37 °C	10 to 35 °C	20 to 37 °C	20 °C
Angular response	Isotropic	Isotropic	Isotropic	Isotropic

Size	145 mm length × 19 mm diameter	120 mm length × 16 mm diameter	145 mm length × 19 mm diameter	80 mm length × 16 mm diameter
Weight	58 g	33 g	58 g	20 g
Re-use	Yes	Yes	Yes	>10 cycles
Warranty	90 d	90 d	90 d	90 d
Other comments		Temperature response curve provided	Thermal:fast neutron sensitivity >10:1	Special recompression chamber available

[a]BDT = bubble detector thermal.

Glossary

absorbed dose (D): The quotient of dE by dm, where dE is the mean energy imparted by ionizing radiation to matter of mass dm. The unit for absorbed dose is the joule per kilogram (J kg^{-1}), with the special name gray (Gy).

accelerator: In this Report, refers to an electron accelerator, which is a device for imparting kinetic energy to electrons, with the kinetic energy being >2 and <50 MeV.

accelerator head: The part of the accelerator enclosing the x-ray target or source from which the useful beam emanates. The accelerator head contains shielding and may rotate about an axis.

activation: The process of inducing radioactivity by irradiation. An example is the process of creating radionuclides by neutron and gamma-ray activation of materials within the treatment room.

activity: The number of spontaneous nuclear transformations that occur in a quantity of a radioactive nuclide per unit time. The unit of activity is one transformation per second (s^{-1}) with the special name becquerel (Bq).

air kerma (K_a): (see **kerma**).

annihilation: The process by which electromagnetic radiation is emitted as a result of the combination and disappearance of an electron and a positron. Two gamma rays of 0.511 MeV energy each are emitted in most cases.

as low as reasonably achievable (ALARA): A principle of radiation protection philosophy that requires that exposures to ionizing radiation be kept as low as reasonably achievable, economic and social factors being taken into account. The protection from radiation exposure is ALARA when the expenditure of further resources would be unwarranted by the reduction in exposure that would be achieved.

atomic number (Z) (low-Z, high-Z): The atomic number of a nucleus is the number of protons contained in the nucleus. Low-Z describes nuclei with $Z \leq 26$. High-Z describes nuclei with $Z > 26$.

attenuation: The reduction of dose equivalent or other physical properties of a radiation field upon the passage of radiation through matter. This Report is concerned primarily with broad-beam attenuation that occurs when the area of the radiation field is large at the barrier (in contrast to a small diameter beam).

barn: Special unit for the cross section. 1 barn = 10^{-28} m^2 (10^{-24} cm^2) (see **cross section**).

barrier (or **protective barrier**): A protective wall of radiation attenuation material(s) used to reduce the dose equivalent on the side beyond the radiation source (see **primary** and **secondary barriers**).

beam-on time: The time that the radiation source is actually producing radiation.

bremsstrahlung: The spectrum of photons produced by the acceleration or deceleration of high-energy electrons, particularly near the coulomb fields of nuclei (see also **x-ray target**).

broad beam: Conditions of a radiation-shielding situation in which the beam impinging on a barrier surface includes scattered radiation and is laterally extensive.

collimator: A device used to reduce the cross-sectional area of the useful beam of photons or electrons with an absorbing material.

concrete: (see **ordinary concrete**).

controlled area: A limited-access area in which the occupational exposure of personnel to radiation or to radioactive material is under the supervision of an individual in charge of radiation protection. This implies that access, occupancy, and working conditions are controlled for radiation protection purposes.

cross section (σ): The quotient of probability by particle fluence, referring to the probability of an interaction for a single target entity when subjected to a given particle fluence [see **fluence (particle)**]. The interaction is produced by incident charged or uncharged particles. The special unit for the cross section is **barn**.

deep dose equivalent: Dose equivalent at a tissue depth of 1 cm (NRC, 1996). Also called personnel dose equivalent at a depth of 1 cm (ICRU, 1993). The unit for deep dose equivalent is J kg^{-1}, with the special name sievert (Sv).

directly ionizing radiation: Charged particles (electrons, protons, alpha particles) having sufficient kinetic energy to produce ionization by collision (see **ionizing radiation**).

direct radiation: Radiation emitted from the target or source that passes though the collimator opening (see also **useful beam** and **primary radiation**).

dose: In this Report, used as a generic term when not referring to a specific quantity, such as absorbed dose or dose equivalent.

dose equivalent (H): The product of absorbed dose (D) and the radiation quality factor at a specified point of interest in tissue. The unit for H is joule per kilogram (J kg^{-1}), with the special name sievert (Sv).

effective dose (E): The sum of the weighted equivalent doses for the radiosensitive tissues and organs of the body. It is given by the expression $E = \Sigma_T (w_T H_T)$, in which H_T is the equivalent dose in tissue or organ T and w_T is the tissue weighting factor for tissue or organ T. The unit of E and H_T is joule per kilogram (J kg^{-1}), with the special name sievert (Sv) (see **equivalent dose** and **tissue weighting factor**).

energy, low or high: In this Report, for bremsstrahlung, low-energy is 10 MV or less, and high-energy is >10 MV. For a particle, unless otherwise specified, the term refers only to its kinetic energy in million electron volts.

equilibrium tenth-value layer (TVL_e): The thickness of a specific material that attenuates a specified radiation by a factor of 10, under broad-beam conditions, in that penetration region in which the directional and spectral distributions of the radiation are practically independent of thickness.

equivalent dose (H_T): The mean absorbed dose (D_T) in a tissue or organ modified by the radiation weighting factor (w_R) for the type and energy of the radiation. The unit for H_T is J kg^{-1}, with the special name sievert (Sv).

exposure: In this Report, exposure is used most often in its general sense, meaning to be irradiated. When used as the specifically defined radiation quantity, exposure is a measure of the ionization produced in air by x or gamma radiation. The unit of exposure is coulomb per kilogram (C kg^{-1}). The special unit for exposure is roentgen (R), where 1 R = 2.58×10^{-4} C kg^{-1}.

fluence (particle) (Φ): The quotient of d*N* by d*a*, where d*N* is the number of particles or photons that enter a sphere of cross-sectional area d*a*. The unit for particle fluence is m^{-2}, but it is also commonly expressed in cm^{-2} (*i.e.*, particles per m^2, or per cm^2).

fluence (particle) rate (φ): The quotient of dΦ by d*t*, where dΦ is the increment of particle fluence in the time interval d*t*. The unit for particle fluence rate is m^{-2} s^{-1}, but it is also commonly expressed in cm^{-2} s^{-1}(*i.e.*, particles per m^2 s^{-1}, or per cm^2 s^{-1}).

gamma ray: A photon emitted in the process of nuclear transition or radioactive decay.

gantry: The rotating arm on which the accelerator head (or ^{60}Co source) is mounted. The gantry, and therefore the useful beam of radiation, typically can rotate 360 degrees about its axis.

gray (Gy): The special name for the unit of the quantities absorbed dose and air kerma. 1 Gy = 1 J kg^{-1}.

half-life, radioactive: The time for the activity of any particular radionuclide to be reduced to one-half its initial value.

half-value layer (*HVL*): The thickness of a specified substance which, when introduced into the path of a given beam of radiation, reduces the radiation field quantity to one-half its original value.

indirectly ionizing radiation: Uncharged particles (*e.g.*, neutrons, photons, gamma rays) that are capable of releasing charged particles when interacting with matter (see **ionizing radiation**).

instantaneous dose-equivalent rate (IDR): The dose-equivalent rate in Sv h^{-1} as measured with the accelerator operating at the absorbed-dose output rate $\dot{D}_0$. *IDR* is specified at 30 cm beyond the penetrated barrier.

interlock: Device that automatically shuts off or reduces the radiation emission rate from an accelerator to acceptable levels (*e.g.*, by the opening of a door into a radiation area). In certain applications, an interlock can be used to prevent entry into a treatment room.

inverse square law: The rule that states that the intensity of radiation from a point source decreases as $1/d^2$ from the source in a nonabsorbing medium, where *d* is the distance from the source.

ionizing radiation: Any radiation consisting of directly or indirectly ionizing particles or photons or a mixture of both. These radiations can produce ions as a consequence of interactions with matter.

irradiation: Exposure to ionizing radiation (see also **exposure**).

isocenter: The point defined by intersection of the gantry axis of rotation and the beam centerline of a medical accelerator or cobalt unit. Typically, the isocenter is located 1 m from the radiation source.

kerma (kinetic energy released per unit mass): The sum of the initial kinetic energies of all the charged particles liberated by uncharged particles per unit mass of specified material. The unit for kerma is J kg^{-1}, with the special name gray (Gy). Kerma can be quoted for any specified material at a point in free space or in an absorbing medium (*e.g.*, air kerma).

lead equivalence: The thickness of lead affording the same attenuation, under specified conditions, as the material in question.

leakage radiation: All radiation, except the useful beam, coming from within the accelerator head and other beam-line components. It is attenuated by shielding in the protective source housing as specified by IEC (2002).

mean absorbed dose (D_T): The mean absorbed dose in an organ or tissue, obtained by integrating or averaging absorbed doses at points in the organ or tissue.

members of the public: All persons who are not already considered occupationally exposed by a source or practice under consideration. When being irradiated as a result of medical care, patients are a separate category.

monitor unit (MU): The unit of measure of the quantity of ionizing radiation passing through a monitor chamber assembly located in the path of the useful beam from an accelerator. The value of the monitor unit is determined by calibrating the resulting absorbed dose in water usually at the isocenter under specified conditions.

monoenergetic: Possessing a single energy, or being within a narrowly-limited band of energies.

narrow beam: Conditions in which the measurement of ionizing radiation passing through a barrier does not include a contribution from scattered radiation within the barrier. These conditions can be met with a parallel beam of radiation having a small cross-sectional area impinging on a thin barrier and using a small detector located far from the barrier.

neutron capture: A process in which a neutron becomes part of the nucleus with which it interacts without release of another heavy particle.

neutron capture gamma ray: A photon emitted as an immediate result of the neutron-capture process.

neutron source strength (Q_n): The number of neutrons emitted from the head of the linear accelerator per gray of x-ray absorbed dose at the isocenter.

occupancy factor (T): The factor ($\leq$1) by which the workload should be multiplied to correct for the degree of occupancy (by any one person) of the area in question while the radiation source is in the "on" position and emitting radiation.

occupational exposure: Exposures to individuals that are incurred in the workplace as the result of situations that can reasonably be regarded as being the responsibility of management (exposures associated with medical diagnosis or treatment are excluded).

occupied area: Any room or other space, indoors or outdoors, that is likely to be occupied by any person, either regularly or periodically during the course of the person's work, habitation or recreation, and in which an ionizing radiation field exists because of radiation sources in the vicinity.

ordinary concrete: A Portland-cement concrete whose constituents are those usually utilized in construction. Thus, ordinary concrete excludes those mixtures called heavy concrete, in which a special material (*e.g.*, iron) has been added to enhance the radiation-shielding properties. Cured ordinary concrete is specified with a density of 2.35 g cm^{-3}. Other terms found in the literature that are synonymous with ordinary concrete are standard-weight and normal-weight concrete. Often the density is rounded off to 2.4 g cm^{-3}.

phantom: As used in this Report, a volume of tissue- or water-equivalent material used to simulate the absorption and scattering characteristics of the patient's body or portion thereof.

photoneutron: A neutron released from a nucleus as the result of the absorption of an energetic photon.

photon: An energy quantum of electromagnetic radiation. In this Report, an x or gamma ray.

point source (of radiation): Any radiation source as viewed from a distance that is much greater than the linear size of the source, and for which the inverse square law is applicable. In this Report, when the distance from the source exceeds 10 times the largest linear dimension of the source, it may be considered a point source.

primary beam: (see **primary radiation**).

primary barrier: A wall, ceiling, floor or other structure designed to attenuate the useful beam to the required degree.

primary radiation (useful beam): In this Report, radiation emitted directly from the source that is intended to be used for medical purposes.

protective source housing: The part of the accelerator or teletherapy unit enclosing the x-ray target and/or source(s) from which the useful beam emanates. This component contains shielding and may rotate about an axis.

pulse cycle: The fraction of the operation cycle of an accelerator during which radiation is produced; the product of the pulse duration and the pulse-repetition frequency.

qualified expert: A medical or health physicist who is competent to design radiation shielding in radiotherapy facilities, and who is certified by the American Board of Radiology, American Board of Medical Physics, American Board of Health Physics, or Canadian College of Physicists in Medicine.

radiation protection survey: An evaluation of the radiation protection in and around an installation that includes radiation measurements, inspections, evaluations and recommendations.

reflection coefficient (α): The fraction of radiation (*e.g.*, fluence, energy, absorbed dose) expressed by the ratio of the amount backscattered to that incident.

scattered radiation: Radiation that, during passage through matter, is changed in direction, and the change is usually accompanied by a decrease in energy.

scatter fraction [$a(\theta)$]: The ratio of absorbed dose of photons at 1 m from a tissue-equivalent scattering object to the absorbed dose measured at the isocenter at the surface of the scattering object with the objects removed. This quantity is a function of the scatter angle (θ), incident beam quality, and beam area. A scattering phantom is typically a water-equivalent sphere or right circular cylinder of 28 to 30 cm diameter.

secondary barrier: A wall, ceiling, floor or other structure designed to attenuate the leakage and scattered radiations to the required degree.

secondary radiation: All radiation produced by scattering off of objects or leakage through the protective source housing of the treatment unit. That is, all radiation in the treatment room except for the primary beam.

shielding design goal (*P*)**:** Practical values, for a single radiotherapy source or set of sources, that are evaluated at a reference point beyond a protective barrier. When used in conjunction with the conservatively safe assumptions recommended in this Report, the shielding design goals will ensure that the respective annual values for effective dose recommended by NCRP (2004) for controlled and uncontrolled areas are not exceeded. For mixtures of low and high linear-energy-transfer radiation, the quantity dose equivalent is used. *P* can be expressed as a weekly or annual value (*e.g.*, mSv week^{-1} or mSv y^{-1} dose equivalent), but is most often expressed as weekly values since the workload for a radiotherapy source has traditionally utilized a weekly format.

skyshine: Radiation scattered back to Earth by the atmosphere above a radiation-producing facility.

slant thickness (t_s)**:** For radiation that is obliquely incident on a shielding barrier, the slant thickness (t_s) equals $t/\cos\,\theta$, where is the angle of obliquity and t is the thickness of the barrier.

slowing down (of neutrons): Decrease in neutron kinetic energy, usually due to repetitive collisions with the matter through which they traverse.

tenth-value layer (*TVL*)**:** The thickness of a specified substance which, when introduced into the path of a given beam of radiation, reduces the radiation field quantity to one-tenth of its original value.

tenth-value distance (*TVD*)**:** The distance that radiation must traverse in order to reduce the radiation field quantity to one-tenth of its original value.

thermal neutrons: Neutrons in thermal equilibrium with their surroundings. In this Report, all neutrons with energies of less than ~1 eV are considered "thermal."

threshold, radiation-effect (or radiation-damage): The minimum absorbed dose (or dose equivalent) of radiation that will produce a specified effect or a specified type of damage to the irradiated material.

time averaged dose-equivalent rate (TADR): The barrier attenuated dose-equivalent rate averaged over a specified time or period of accelerator operation. TADR is proportional to instantaneous dose-equivalent rate (*IDR*), and depends on the values of workload (*W*) and use factor (*U*).

transmission factor (or barrier transmission) (for photons or neutrons) (*B*)**:** For a given radiation type and quality, *B* is the ratio of any radiation field quantity at a location behind the barrier on which radiation is incident to the field quantity at the same location without the presence of the shield. *B* is a measure of the shielding effectiveness of the barrier.

two-source rule: This phrase refers to the conservatively safe, and often used, guideline that, when a location is to be shielded from two different sources of radiation, each passing through the same barrier, the resultant thickness of the barrier should be equal to the greater of the two individual thicknesses if they differ from one another by more than a *TVL*, or else it should be equal to the greater thickness plus one added *HVL*, as determined by the more penetrating of the two radiation sources.

uncontrolled area: Any space not meeting the definition of controlled area.

useful beam: (see **primary radiation**).

vault: A shielded room in which a high-intensity radiation source is housed.

week, calendar: Seven consecutive days.

week, work: Any combination of time intervals adding up to 40 h within seven consecutive days.

workload (W): The average absorbed dose of radiation produced by a source over a specified time (most often one week) at a specific location. In this Report, the workload is defined as the absorbed dose from photons at the isocenter, at 1 m from the source over a one week period averaged over a year. This Report defines two workload quantities: primary workload, and leakage-radiation workload.

workload, primary (W_{pri}): The workload arising from the primary beam (or useful beam).

workload, leakage-radiation (W_L): The workload arising from leakage radiation and measured at 1 m from the source of leakage radiation.

x-ray target: In this Report, the high-atomic number material used to convert an energetic electron beam into x rays by the bremsstrahlung process.

Symbols and Acronyms

α	reflection coefficient
β	transmission factor for neutrons that penetrate the head shielding
θ	angle for patient-scattered radiation
σ	cross section for thermal neutron reaction
φ_A	total neutron fluence at a point per unit absorbed dose of x rays at the isocenter (n m^{-2} Gy^{-1})
$\Phi(E)$	neutron fluence as a function of energy (n cm^{-2})
$\dot{\Phi}(E)$	neutron-fluence rate at 1 m from the target (n cm^{-2} h^{-1})
Ω	solid angle of the maximum beam
A	atomic mass number
A_0	beam area at the first scattering surface
A_1	area of wall that can be seen from maze door
A_z	cross-sectional area of maze inner entry projected onto the maze wall from the perspective of the irradiated primary barrier
$a(\theta)$	scatter fraction or fraction of the primary-beam absorbed dose that scatters from the patient at a particular angle
ALARA	as low as reasonably achievable
B	transmission factor or barrier transmission
B_{conc}	transmission factor for concrete
BD-100R	bubble detector, threshold = 100 keV, reusable
BD-PND	bubble detector, personal neutron dosimeter
BDS	bubble detector spectrometer
BDT	bubble detector thermal
B_L	barrier transmission of leakage radiation
B_{Pb}	transmission factor for lead
BPE	borated polyethylene
B_{pri}	transmission factor of the primary barrier
B_{ps}	barrier transmission for radiation scattered by the patient
B_{steel}	transmission factor for steel
B_T	total transmission factor for a barrier
B_{xs}	roof shielding transmission factor for photons
C	calibration constant (sievert per count)
C_{corr}	corrected number of counts
C_I	IMRT factor (ratio of MU_{IMRT} to MU_{conv})
C_n	measured number of counts
C_{QA}	quality assurance factor
d	distance (subscripted for specific distance described)[29]
D	absorbed dose

[29]See also Figure 7.1 for additional distance notations used in the Section 7 example calculations, but not listed here.

d_0	a distance of 1.41 m, used in an application of Kersey's method
d_h	perpendicular distance from the target to the first reflection surface
d_{hd}	accelerator head-to-maze door distance
d_i	vertical distance from the target to a point 2 m above the roof
d_L	distance from the isocenter (or closest approach of the accelerator head) to the point protected for leakage radiation
d_{LS}	distance from x-ray target to the maze centerline in area A_1
D_o	x-ray absorbed dose per week at isocenter (cGy week^{-1})
$\dot{D}_0$	absorbed-dose output rate at 1 m from the target (Gy h^{-1})
$\dot{D}_0$ (MU)	accelerator production rate (MU h^{-1})
d_{pp}	perpendicular distance from isocenter to the wall
D_{pre}	unit prescribed absorbed dose per fraction
d_{pri}	distance from the x-ray target to the point protected
d_r	distance from beam center at the first reflection, past the edge of the inner wall, to Point b on the mid-line of the maze
d_s	distance from the isocenter
d_{sca}	distance from the x-ray target to the patient or scattering surface
d_{sec}	distance from the scattering object to the point protected
D_T	mean absorbed dose in a tissue or organ
d_{TBI}	total-body irradiation treatment distance
D_{TBI}	total weekly absorbed dose to the patient for total-body irradiation
d_z	centerline distance along the maze from Point b to the maze door
d_{zz}	centerline distance along the maze length from a scattering surface to the door
E	effective dose
E_n	neutron energy
$\bar{E}_n$	average neutron energy
E_Q	kinetic energy released for a thermal neutron reaction
E_{th}	threshold energy for a photon-induced effect
f	fraction of the primary beam transmitted through the patient
$f(\theta)$	angular distribution of the roof-scattered photons
F	field area at the mid-depth of the patient at 1 m
F_{max}	maximum field area at isocenter
G	gantry angle
H	dose equivalent
H_0	total (direct plus room-scattered plus thermal) neutron dose equivalent at a distance d_0 from the target per unit absorbed dose of x rays at the isocenter
$H^*(10)$	ambient dose equivalent at a depth of 10 mm
$\dot{H}$	dose-equivalent rate
h_φ	dose equivalent from neutron capture gamma rays at the outside maze entrance per unit absorbed dose of x rays at the isocenter
$h_\Phi(E)$	fluence to ambient-dose-equivalent conversion function

H_{cg}	weekly dose equivalent at maze door due to neutron capture gamma rays
H_G	total dose equivalent at the maze door
H_{LS}	dose equivalent per week due to head-leakage photons scattered by the room surfaces
H_{LT}	dose equivalent per week due to leakage radiation that is transmitted through the inner maze wall.
H_{MU}	dose equivalent per monitor unit
$H_{MU,L}$	leakage-radiation dose equivalent per monitor unit (without the phantom and with the collimator closed)
$H_{MU,ps}$	phantom-scattered-radiation dose equivalent per monitor unit
$H_{MU,total}$	total short-term measured dose equivalent per monitor unit (made with phantom in the beam)
H_n	neutron dose equivalent per week
$\dot{H}_n$	neutron dose-equivalent rate
$H_{n,D}$	neutron dose equivalent at maze entrance per unit absorbed dose of x rays at the isocenter
H_{ns}	ratio of the dose equivalent beyond the ceiling shield to the neutron fluence incident at the ceiling
H_p	dose equivalent from patient-scattered radiation at the maze entrance
H_{phtr}	photon dose equivalent
H_{ps}	dose equivalent per week at the maze door due to patient-scattered radiation
$H_{ps,iso}$	dose equivalent from radiation scattered by patient, isocentric techniques
$H_{ps,TBI}$	dose equivalent from radiation scattered by the patient, TBI techniques
$\overline{H}_{pt}$	average dose equivalent per patient treatment at 30 cm beyond the penetrated barrier
H_S	dose equivalent per week due to scatter of the primary beam from the room surfaces
H_{sec}	total dose equivalent due to scattered and leakage radiations
$\dot{H}_{SS}$	side-scattered dose-equivalent rate
H_T	equivalent dose to a specific organ or tissue
H_{Tot}	total dose equivalent beyond a barrier
H_{tr}	calculated transmitted x-ray dose equivalent
H_W	total weekly dose equivalent at external maze entrance (leakage and scattered radiations, neutron capture gamma rays, and neutrons)
HVAC	heating, ventilation and air conditioning
HVL	half-value layer
IDR	instantaneous dose-equivalent rate with the accelerator operating at maximum output at 30 cm beyond a barrier
IDR_L	instantaneous dose-equivalent rate measured at a point located 30 cm beyond the secondary barrier in the absence of a phantom at the isocenter
IDR_{ps}	instantaneous dose-equivalent rate at a point 30 cm beyond the secondary barrier due to patient-scattered radiation

IDR_{total}	instantaneous dose-equivalent rate measured at a point located 30 cm beyond the secondary barrier in the presence of a phantom
IMRT	intensity modulated radiation therapy
IORT	intraoperative radiotherapy
K	ratio of the neutron capture gamma-ray dose equivalent to the total neutron fluence
K_a	air kerma
L_f	head leakage radiation ratio at 1 m from the target
LET	linear-energy transfer
M	maximum number of patient treatments in-any-one-hour divided by the average number of patient treatments per hour
MU	monitor units
MU_{conv}	monitor unit per unit absorbed dose for conventional treatment
MU_{IMRT}	average monitor unit per unit prescribed absorbed dose needed for intensity modulated radiation therapy
n	number of tenth-value layers
N_h	average number of patient treatments per hour
N_{max}	maximum number of patient treatments in-any-one-hour
N_w	average number of patient treatments per week
O_3	ozone
P	shielding design goal
P_r	pulse repletion rate (pulses per second)
QA	quality assurance
Q_n	neutron source strength in neutrons emitted from the accelerator head per gray of x-ray absorbed dose at the isocenter
$r_{\Phi}(E)$	response function of rem-meter (counts per unit fluence)
R	neutron production coefficient
R_h	time averaged dose-equivalent rate, in-any-one-hour
R_m	rem-meter response
R_w	weekly time averaged dose-equivalent rate
S_0	inner maze entrance cross-sectional area
S_1	cross-sectional area along the maze
S_r	surface area of the treatment room
SRS	stereotactic radiosurgery
SRT	stereotactic radiotherapy
t	barrier thickness
t_1	first concrete slab thickness
t_2	second concrete slab thickness
t_s	slant thickness
T	occupancy factor
TADR	time averaged dose-equivalent rate
$t_{barrier}$	barrier thickness
TBI	total-body irradiation
T_D	dead time
TLD	thermoluminescent dosimeter
T_p	pulse width
TVD	tenth-value distance
TVL	tenth-value layer

TVL_1	first tenth-value layer
TVL_e	equilibrium tenth-value layer
TVL_n	tenth-value layer in concrete for neutrons
TVL_x	tenth-value layer in concrete for the primary x-ray beam
U	use factor
U_G	use factor for the Wall G or for the gantry orientation G
$U(G)$	use factor as a function of the gantry angle
U_{pri}	use factor for the primary barrier
U_{ps}	use factor for the gantry orientation used during the measurements
U_x	use factor or fraction of time that beam is likely to be incident on the barrier for procedure type x
v	velocity
W	workload for radiotherapy equipment (absorbed dose delivered to the isocenter in a week) (Gy $week^{-1}$)
W_{conv}	total workload (at the isocenter) for conventional techniques (Gy $week^{-1}$)
W_{IMRT}	total workload (at the isocenter) for IMRT techniques (Gy $week^{-1}$)
W_L	workload for leakage radiation (Gy $week^{-1}$)
$W_L(MU)$	leakage-radiation workload (MU $week^{-1}$)
$W(MU)$	primary-beam workload (MU $week^{-1}$)
W_{pri}	primary-barrier weekly workload (Gy $week^{-1}$)
W_{ps}	patient-scattered radiation workload (Gy $week^{-1}$)
$W_{ps}(MU)$	patient-scattered radiation workload (MU $week^{-1}$)
W_{QA}	total workload (at the isocenter) for quality assurance techniques (Gy $week^{-1}$)
w_R	radiation weighting factor
w_T	tissue weighting factor for an organ or tissue
W_{TBI}	total-body irradiation workload (absorbed dose at 1 m) (Gy $week^{-1}$)
$WU]_{pri}$	workload-use factor product for the primary barrier
$WU]_{wallscat}$	workload-use factor product for wall-scattered radiation
W_x	workload at 1 m for procedure type x (Gy $week^{-1}$)
X_R	distance from the beam center at the rooftop to the point of interest
Z	atomic number

References

AAPM (1983). American Association of Physicists in Medicine, Task Group 21, Radiation Therapy Committee. "A protocol for the determination of absorbed dose from high-energy photon and electron beams," Med. Phys. **10**(6), 741-771.

AAPM (1986a). American Association of Physicists in Medicine. *The Physical Aspects of Total and a Half Body Photon Irradiation*, AAPM Report No. 17 (Medical Physics Publishing, Madison, Wisconsin).

AAPM (1986b). American Association of Physicists in Medicine. *Neutron Measurements Around High Energy X-Ray Radiotherapy Machines*, AAPM Report No. 19 (Medical Physics Publishing, Madison, Wisconsin).

AAPM (1995). American Association of Physicists in Medicine. *Stereotactic Radiosurgery*, AAPM Report No. 54 (Medical Physics Publishing, Madison, Wisconsin).

ABRATH, F.G., BELLO, J. and PURDY, J.A. (1983). "Attenuation of primary and scatter radiation in concrete and steel for 18 MV x-rays from a Clinac-20 linear accelerator," Health Phys. **45**(5), 969–973.

ACR (2000). American College of Radiology. "Worker safety in radiation therapy suites: Automatic door systems may pose danger," ACR Newsletter **56**(6), 11.

ALMEN, A., AHLGREN, L. and MATTSSON, S. (1991). "Absorbed dose to technicians due to induced activity in linear accelerators for radiation therapy," Phys. Med. Biol. **36**, 815–822.

AL-AFFAN, I.A. (2000). "Estimation of the dose at the maze entrance for x-rays from radiotherapy linear accelerators," Med. Phys. **27**(1), 231–238.

ANDERSSON, I.O. and BRAUN, J.A. (1963). "A neutron rem-counter with uniform sensitivity from 0.025 eV to 10 MeV," pages 87 to 95 *Neutron Dosimetry, Volume II*, IAEA Proceedings Series, STI/PUB/69 (International Atomic Energy Agency, Vienna).

AXTON, J. and BARDELL, A.G. (1979). "Neutron production from electron accelerators used for medical purposes," in *Proceedings of a Conference on Neutrons from Electron Medical Accelerators*, NBS Special Publication 554, Heaton, H.T. and Jacobs, R., Eds. (U.S. Government Printing Office, Washington).

BARISH, R.J. (1993). "Evaluation of a new high-density shielding material" Health Phys. **64**(4), 412–416.

BARISH, R.J. (2005). "Minimizing entrance door thickness for direct-entry radiotherapy rooms," Health Phys. **89**(2), 168–171.

BARTLETT, D.T., TANNER, R.J., TAGZIRIA, H. and THOMAS, D.J. (2002). *Response Characteristics of Neutron Survey Instruments*, NRPB-R333(rev) (National Radiological Protection Board, Chilton, Didcot, Oxon, United Kingdom).

BIGGS, P. (1996). "Obliquity factors for ^{60}Co and 4, 10, and 18 MV x rays for concrete, steel, and lead and angles of incidence between 0 and 70 degrees," Health Phys. **70**(4), 527–536.

BRAMBLETT, R.L., EWING, R.I. and BONNER, T.W. (1960). "A new type of neutron spectrometer," Nucl. Instrum. Methods **9**, 1–12.

BTI (2005). Bubble Technology Industries. http://www.bubbletech.ca/b_spec.htm (accessed December 2005) (Bubble Technology Industries, Chalk River, Ontario, Canada).

COHEN, M., Ed. (1972). "Central axis depth dose data for use in radiotherapy," Brit. J. Radiol. **11** (Suppl. 11).

COSACK, M. and LESIECKI, H. (1985). "Dose equivalent survey meters," Radiat. Prot. Dosim. **10**(1–4), 111–119.

CRCPD (1991). Conference of Radiation Control Program Directors, Inc. "Radiation safety requirements for particle accelerators," Part I in *Suggested State Regulations for Control of Radiation: Volume I, Ionizing* (Conference of Radiation Control Program Directors, Inc., Frankfort, Kentucky).

CRCPD (1999). Conference of Radiation Control Program Directors, Inc. "Therapeutic radiation machines," Part X in *Suggested State Regulations for Control of Radiation: Volume I, Ionizing* (Conference of Radiation Control Program Directors, Inc., Frankfort, Kentucky).

DAVES, J.L. and MILLS, M.D. (2001). "Shielding assessment of a mobile electron accelerator for intraoperative radiotherapy," J. Appl. Clin. Med. Phys. **2**(3), 165–173.

DEYE, J.A. and YOUNG, F.C. (1977). "Neutron production from a 10 MV medical linac," Phys. Med. Biol. **22**(1), 90–94.

DOE (1993). U.S. Department of Energy. "Occupational radiation protection," 10 CFR 835 (U.S. Government Printing Office, Washington).

FOLLOWILL, D., GEIS, P. and BOYER, A. (1997). "Estimates of whole-body dose equivalent produced by beam intensity modulated conformal therapy," Int. J. Radiat. Oncol. Biol. Phys. **38**(3), 667–672 (Errata **38**, 783).

FOLLOWILL, D.S., STOVALL, M.S., KRY, S.F. and IBBOTT, G.S. (2003). "Neutron source strength measurements for Varian, Siemens, Elekta, and General Electric linear accelerators," J. Appl. Clin. Med. Phys. **4**(3), 189–194.

HANKINS, D.E. and CORTEZ, J.R. (1974). *Directional Response and Energy Dependence of Four Neutron Remmeters*, LA-5528 (Los Alamos National Laboratory, Los Alamos, New Mexico).

HANSON, A.O. and MCKIBBEN, J.L. (1947). "A neutron detector having uniform sensitivity from 10 keV to 3 MeV, Phys. Rev. **72**, 673–677.

IAEA (1979). International Atomic Energy Agency. *Radiological Safety Aspects of the Operation of Electron Linear Accelerators*, Technical Reports Series No. 188 (International Atomic Energy Agency, Vienna).

ICRP (1964). International Commission on Radiological Protection. *Report of Committee IV on Protection Against Electromagnetic Radiation Above 3 MeV and Electrons, Neutrons and Protons*, ICRP Publication 4 (Elsevier Science, New York).

ICRP (1973). International Commission on Radiological Protection. *Data for Protection Against Ionizing Radiation from External Sources: Supplement to ICRP Publication 15*, ICRP Publication 21 (Elsevier Science, New York).

ICRP (1987). International Commission on Radiological Protection. *Data for Use in Protection Against External Radiation*, ICRP Publication 51, Ann. ICRP **17**(2-3) (Elsevier Science, New York).

ICRP (1991). International Commission on Radiological Protection. *1990 Recommendations of the International Commission on Radiological Protection*, ICRP Publication 60, Ann. ICRP **21**(1-3) (Elsevier Science, New York).

ICRP (1996). International Commission on Radiological Protection. *Conversion Coefficients for Use in Radiological Protection Against External Radiation*, ICRP Publication 74, Ann. ICRP **26**(3) (Elsevier Science, New York).

ICRU (1993). International Commission on Radiation Units and Measurements. *Quantities and Units in Radiation Protection Dosimetry*, ICRU Report 51 (Oxford University Press, Cary, North Carolina).

ICRU (1998). International Commission on Radiation Units and Measurements. *Conversion Coefficients for Use in Radiological Protection Against External Radiation*, ICRU Report 57 (Oxford University Press, Cary, North Carolina).

IEC (2002). International Electrotechnical Commission. *Amendment 1. Medical electrical equipment – Part 2-1: Particular requirements for the safety of electron accelerators in the range 1 MeV to 50 MeV*, IEC 60601-2-1-AM1, 2nd ed. (International Electrotechnical Commission, Geneva).

ING, H., NOULTY, R.A., and MCLEAN, T.D. (1997). "Bubble detectors – a maturing technology," Radiat. Meas. **27**, 1–11.

IPE, N.E. and. BUSICK, D.D (1987). *BD–100: The Chalk River Nuclear Laboratories' Neutron Bubble Detector*, SLAC-PUB-4398 (Stanford Linear Accelerator Center, Stanford, California).

IPE, N.E., BUSICK, D.D. and POLLOCK, R.W. (1988). "Factors affecting the response of the bubble detector BD-100 and a comparison of its response to CR-39," Radiat. Prot. Dosim. **23**, 135–138.

IPE, N.E., ROESLER, S., JIANG, S.B. and MA, C.M. (2000). *Neutron Measurements for Intensity Modulated Radiation Therapy*, SLAC-PUB-8443 (Stanford Linear Accelerator Center, Stanford, California).

KASE, K.R., MAO, X.S., NELSON, W.R., LIU, J.C., KLECK, J.H. and ELSALIM, M. (1998). "Neutron fluence and energy spectra around the Varian Clinac 2100C/2300C medical accelerator," Health Phys. **74**(1), 38–47.

KASE, K.R., NELSON, W.R., FASSO, A., LIU, J.C., MAO, X, JENKINS, T.M. and KLECK, J.H. (2003). "Measurements of accelerator-produced leakage neutron and photon transmission through concrete," Health Phys. **84**(2), 180–187.

KERSEY, R.W. (1979). "Estimation of neutron and gamma radiation doses in the entrance mazes of SL75-20 linear accelerator treatment rooms," Medicamundi **24**, 151–155.

KIRN, F.S. and KENNEDY, R.J. (1954). "How much concrete for shielding: Betatron x-rays," Nucleonics **12**(6), 44.

KIRN, F.S., KENNEDY, R.J. and WYCKOFF, H.O. (1954). "The attenuation of gamma rays at oblique incidence," Radiology **63**(1), 94–104.

KLECK, J.H. and ELSALIM, M. (1994). "Clinical workloads and use factors for medical linear accelerators," (abstract) Med. Phys. **21**, 952–953.

KNOLL, G.F. (1989). *Radiation Detection and Measurement*, 2nd ed. (John Wiley and Sons, New York).

LALONDE, R. (1997). "The effect of neutron-moderating materials in high-energy linear accelerator mazes," Phys. Med. Biol. **42**, 335–344.

LARIVIERE, P.D. (1985). "Radiotherapy technologist dose from high-energy electron medical accelerators," Health Phys. **49**(6), 1105–1114.

LO, Y.C. (1992). "Albedos for 4-, 10-, and 18-MV bremsstrahlung x-ray beams on concrete, iron, and lead--normally incident," Med. Phys. **19**(3), 659–666.

MAERKER, R.E. and MUCKENTHALER, F.J. (1967). "Neutron fluxes in concrete ducts arising from incident epicadmium neutrons: Calculations and experiment," Nucl. Sci. Eng. **30**, 340.

MACKIE, T.R., HOLMES, T., SWERDLOFF, S., RECKWERDT, P., DEASY, J.O., YANG, J., PALIWAL, B. and KINSELLA, T. (1993). "Tomotherapy: A new concept for the delivery of dynamic conformal radiotherapy," Med. Phys. **20**(6), 1709–1719.

MAO, X.S., KASE, K.R., LIU, J.C., NELSON, W.R., KLECK, J.H. and JOHNSEN, S. (1997). "Neutron sources in the Varian Clinac 2100C/2300C medical accelerator calculated by the EGS4 code," Health Phys. **72**(4), 524–529.

MCCALL, R.C. (1981). *Neutron Measurements*, SLAC-PUB-2662 (Stanford Linear Accelerator Center, Stanford, California)

MCCALL, R.C. (1997). "Shielding for thermal neutrons," Med. Phys. **24**(1), 135–136.

MCCALL, R.C. and KLECK, J.H. (1994). "Neutron shielding for linac primary barriers with steel or lead plus concrete," (abstract) Med. Phys. **21**, 975.

MCCALL, R.C. and SWANSON, W.P. (1979). "Neutron sources and their characteristics," *Proceedings of a Conference on Neutrons from Electron Medical Accelerators*, NBS Special Publication 554, Heaton, H.T. and Jacobs, R., Eds. (U.S. Government Printing Office, Washington).

MCCALL, R.C., JENKINS, T.M. and TOCHILIN, E. (1976). *High Energy Photon Response of Moderated Neutron Detectors*, SLAC-PUB-1768 (Stanford Linear Accelerator Center, Stanford, California).

MCCALL, R.C., JENKINS, T.M. and SHORE, R.A. (1978). *Transport of Accelerator Produced Neutrons in a Concrete Room*, SLAC-PUB-2214 (Stanford Linear Accelerator Center, Stanford, California).

MCCALL, R.C., MCGINLEY, P.H. and HUFFMAN, K.E. (1999). "Room scattered neutrons," Med. Phys. **26**(2), 206–207.

MCDONALD, J.C., SCHWARTZ, R.B. and THOMAS, R.H. (1998). "Neutron dose equivalent conversion coefficients have changed in the last forty years...Haven't they?" Radiat. Prot. Dosim. **78**(2), 147–149.

MCGINLEY, P.H. (1992a). "Photoneutron production in the primary barriers of medical accelerator rooms," Health Phys. **62**(4) 359–362 [Errata **63**(3), 366].

MCGINLEY, P.H. (1992b). "Photoneutron fields in medical accelerator rooms with primary barriers constructed of concrete and metals," Health Phys. **63**(6), 698–701.

MCGINLEY, P.H. (1993). "Radiation skyshine produced by an 18-MeV medical accelerator," Radiat. Prot. Manage. **10**, 59–64.

MCGINLEY, P.H. (2001a) "Direct shielded doors" RSO Magazine **6**(5), 11–19.

MCGINLEY, P.H. (2001b). "Dose rate outside primary barriers," Health Phys. **80** (Suppl. 2), S7–S8.

MCGINLEY, P.H. (2002). *Shielding Techniques for Radiation Oncology Facilities*, 2nd ed. (Medical Physics Publishing, Madison, Wisconsin).

MCGINLEY, P.H. and BUTKER, E.K. (1991). "Evaluation of neutron dose equivalent levels at the maze entrance of medical accelerator treatment rooms," Med Phys. **18**(2), 279–281.

MCGINLEY, P.H. and BUTKER, E.K. (1994). "Laminated primary ceiling barriers for medical accelerator rooms," Phys. Med. Biol. **39**, 1331–1336.

MCGINLEY, P. and HUFFMAN, K.E. (2000). "Photon and neutron dose equivalent in the maze of a high-energy medical accelerator facility" Radiat. Prot. Manage. **17**, 43-46 (Errata **17**, 4).

MCGINLEY, P.H. and JAMES, J.L. (1997). "Maze design methods for 6- and 10-MeV accelerators," Radiat. Prot. Manage. **14**(1), 59–64.

MCGINLEY, P.H. and MINER, M.S. (1995). "A method of eliminating the maze door of medical accelerator rooms," Radiat. Prot. Manage. **12**(5), 29–37.

MCGINLEY, P.H., WRIGHT, B.A. and MEDING, C.J. (1984). "Dose to radiotherapy technologists from air activation" Med. Phys. **11,** 855–858.

MCGINLEY, P.H., LONG, K. and KAPLAN, R. (1988). "Production of photoneutrons in a lead shield by high-energy x-rays," Phys. Med. Biol. **33**, 975–980.

MCGINLEY, P.H., MINER, M.S. and MITCHUM, M.L. (1995). "A method for calculating the dose due to capture gamma rays in accelerator mazes," Phys. Med. Biol. **40**(9), 1467–1473.

MCGINLEY, P.H., DHABA'AN, A.H. and REFT, C.S. (2000). "Evaluation of the contribution of capture gamma rays, x-ray leakage, and scatter to the photon dose at the maze door for a high energy medical electron accelerator using a Monte Carlo particle transport code," Med. Phys. **27**(1), 225–230.

MECHALAKOS, J., ST. GERMAIN, J. and BURMAN, C.M. (2004). "Results of a one year survey of output for linear accelerators using IMRT and non-IMRT techniques," J. Appl. Clin. Med. Phys. **5**(1), 64–72.

MUTIC, S. and LOW, D.A. (1998). "Whole-body dose from tomotherapy delivery," Int. J. Radiat. Oncol. Biol. Phys. **42**(1), 229–232.

MUTSCHELLER, A. (1925). "Physical standards of protection against roentgen ray dangers," Am. J. Roentgenol. **13**, 65–70.

MUTSCHELLER, A. (1926). "Further studies on physical standards of protection against roentgen-ray dangers," Radiology **6**, 314–319.

NATH, R., PRICE, K.W. and HOLEMAN, H.R. (1979). "Mixed photon-neutron field measurements," in *Proceedings of a Conference on Neutrons from Electron Medical Accelerators*, NBS Publication 554, Heaton, H.T. and Jacobs, R., Eds. (U.S. Government Printing Office, Washington).

NBS (1960). National Bureau of Standards. *Measurement of Neutron Flux and Spectra for Physics and Biological Applications*, NBS Handbook No. 72, NCRP Report No. 23 (National Council on Radiation Protection and Measurements, Bethesda, Maryland).

NBS (1964). National Bureau of Standards. *Shielding for High Energy Electron Accelerator Installations*, NBS Handbook No. 97, NCRP

Report No. 31 (National Council on Radiation Protection and Measurements, Bethesda, Maryland).

NBS (1982). National Bureau of Standards. *Medical Physics Data Book*, NBS Handbook No. 138, Padikal, T.N. and Fivozinsky, S.P., Eds. (U.S. Government Printing Office, Washington).

NCRP (1971). National Council on Radiation Protection and Measurements. *Protection Against Neutron Radiation*, NCRP Report No. 38 (National Council on Radiation Protection and Measurements, Bethesda, Maryland).

NCRP (1976). National Council on Radiation Protection and Measurements. *Structural Shielding Design and Evaluation for Medical Use of X Rays and Gamma Rays of Energies up to 10 MeV*, NCRP Report No. 49 (National Council on Radiation Protection and Measurements, Bethesda, Maryland).

NCRP (1977). National Council on Radiation Protection and Measurements. *Radiation Protection Design Guidelines for 0.1-100 MeV Particle Accelerator Facilities*, NCRP Report No. 51 (National Council on Radiation Protection and Measurements, Bethesda, Maryland).

NCRP (1984). National Council on Radiation Protection and Measurements. *Neutron Contamination from Medical Electron Accelerators*, NCRP Report No. 79 (National Council on Radiation Protection and Measurements, Bethesda, Maryland).

NCRP (1985). National Council on Radiation Protection and Measurements. *SI Units in Radiation Protection and Measurements*, NCRP Report No. 82 (National Council on Radiation Protection and Measurements, Bethesda, Maryland).

NCRP (1989). National Council on Radiation Protection and Measurements. *Medical X-Ray, Electron Beam and Gamma-Ray Protection for Energies Up to 50 MeV (Equipment Design, Performance and Use)*, NCRP Report No. 102 (National Council on Radiation Protection and Measurements, Bethesda, Maryland).

NCRP (1990). National Council on Radiation Protection and Measurements. *Implementation of the Principle of as Low as Reasonably Achievable (ALARA) for Medical and Dental Personnel*, NCRP Report No. 107 (National Council on Radiation Protection and Measurements, Bethesda, Maryland).

NCRP (1991). National Council on Radiation Protection and Measurements. *Calibration of Survey Instruments Used in Radiation Protection for the Assessment of Ionizing Radiation Fields and Radioactive Surface Contamination*, NCRP Report No. 112 (National Council on Radiation Protection and Measurements, Bethesda, Maryland).

NCRP (1993). National Council on Radiation Protection and Measurements. *Limitation of Exposure to Ionizing Radiation*, NCRP Report No. 116 (National Council on Radiation Protection and Measurements, Bethesda, Maryland).

NCRP (2003). National Council on Radiation Protection and Measurements. *Radiation Protection for Particle Accelerator Facilities*, NCRP Report No. 144 (National Council on Radiation Protection and Measurements, Bethesda, Maryland).

NCRP (2004). National Council on Radiation Protection and Measurements. *Structural Shielding Design for Medical X-Ray Imaging*

Facilities, NCRP Report No. 147 (National Council on Radiation Protection and Measurements, Bethesda, Maryland).

NELSON, W.R. and LARIVIERE, P.D. (1984). "Primary and leakage radiation calculations at 6, 10 and 25 MeV," Health Phys. **47**(6), 811–818.

NOGUEIRA, I.P. and BIGGS, P.J. (2002). "Measurement of TVLs in lead for 4, 6 and 10 MV bremsstrahlung x-ray beams at scattering angles between 30° and 135°," Health Phys. **83**(2), 255-260.

NRC (1996). U.S. Nuclear Regulatory Commission. "Standards for protection against radiation," 10 CFR 20 (U.S. Government Printing Office, Washington).

NRC (2005a). U.S. Nuclear Regulatory Commission. "Dose limits for individual members of the public," 10 CFR 20.1301(a)(2) (revised January 1), http://www.gpoaccess.gov/cfr/retrieve.html (accessed December 2005) (U.S. Government Printing Office, Washington).

NRC (2005b). U.S. Nuclear Regulatory Commission. "Standards for protection against radiation," 10 CFR 20 (revised January 1), http://www.gpoaccess.gov/cfr/retrieve.html (accessed December 2005) (U.S. Government Printing Office, Washington).

NRC (2005c). U.S. Nuclear Regulatory Commission. "Medical use of byproduct material," 10 CFR 35 (revised January 1), http://www.gpoaccess.gov/cfr/retrieve.html (accessed December 2005) (U.S. Government Printing Office, Washington).

NRC (2005d). U.S. Nuclear Regulatory Commission. "Definitions," 10 CFR 20.1003 (revised January 1), http://www.gpoaccess.gov/cfr/retrieve.html (accessed December 2005) (U.S. Government Printing Office, Washington).

NRC (2005e). U.S. Nuclear Regulatory Commission. "Control of access to high radiation areas," 10 CFR 20.1601 (revised January 1), http://www.gpoaccess.gov/cfr/retrieve.html (accessed December 2005) (U.S. Government Printing Office, Washington).

NRC (2005f). U.S. Nuclear Regulatory Commission. "Control of access to very high radiation areas," 10 CFR 20.1602 (revised January 1), http://www.gpoaccess.gov/cfr/retrieve.html (accessed December 2005) (U.S. Government Printing Office, Washington).

NUMARK, N.J. and KASE, K.R. (1985). "Radiation transmission and scattering for medical linacs producing x rays of 6 and 15 MV: Comparison of calculations with measurements," Health Phys. **48**(3), 289–295.

O'BRIEN, P., MICHAELS, H.B., GILLIES, B., ALDRICH, J.E. and ANDREW, J.W. (1985). "Radiation protection aspects of a new high-energy linear accelerator," Med. Phys. **12**(1), 101–107.

OLSHER, R.H., HSU, H.H., BEVERDING, A., KLECK, J.H., CASSON, W.H., VASILIK, D.G. and DEVINE, R.T. (2000). "WENDI: An improved neutron rem meter," Health Phys. **79**(2), 170–181.

ONGARO, C., ZANINI, A., NATASIS, U., RODENAS, J., OTTAVIANO, G., MANFREDOTTI, C. and BURN, K.W. (2000). "Analysis of photoneutron spectra produced in medical accelerators," Phys. Med. Biol. **45**(12), L55–L61.

PALTA, J.R., BIGGS, P.J., HAZLE, J.D., HUQ, M.S., DAHL, R.A., OCHRAN, T.G., SOEN, J., DOBELBOWER, R.R. JR. and MCCULLOUGH, E.C. (1995). "Intraoperative electron beam radiation therapy: Technique, dosimetry, and dose specification: Report of Task Force 48 of the Radiation Therapy Committee, American Association of

Physicists in Medicine," Int. J. Radiat. Oncol. Biol. Phys. **33**(3), 725–746.

PATTERSON, H.W. and THOMAS, R.H. (1973). *Accelerator Health Physics* (Academic Press, New York).

PROFIO, A. E. (1979). *Radiation Shielding and Dosimetry* (John Wiley and Sons, New York).

PURDY, J.A., BOYER, A.L., BUTLER, E.B., DIPETRILLO, T.A., ENGLER, M.J., FRAASS, B., GRANT, W., III., LING, C.C., LOW, D.A., MACKIE, T.R., MOHAN, R., ROACH, M., ROSENMAN, J.G., VERHEY, L.J., WONG, J.W., CUMBERLIN, R.L., STONE, H. and PALTA, J.R. (2001). "Intensity-modulated radiotherapy: Current status and issues of interest," Int. J. Radiat. Oncol. Biol. Phys. **51**(4), 880–914.

RAWLINSON, J.A., ISLAM, M.K. and GALBRAITH, D.M. (2002). "Dose to radiation therapists from activation at high-energy accelerators used for conventional and intensity-modulated radiation therapy," Med. Phys. **29**(4), 598–608.

ROBINSON, D., SCRIMGER, J.W., FIELD, G.C. and FALLONE, B.G. (2000). "Shielding considerations for tomotherapy," Med. Phys. **27**(10), 2380–2384.

RODGERS, J.E. (2001). "Radiation therapy vault shielding calculational methods when IMRT and TBI procedures contribute," J. Appl. Clin. Med. Phys. **2**(3), 157–164.

RODGERS, J.E. (2005). "CyberKnife treatment room design and radiation protection," pages 41 to 50 in *Robotic Radiosurgery, Vol. 1*, Mould, R.F., Schulz, R.A., Bucholz, R.D., Gagnon, G.J., Gerszten, P.C., Kresl, J.J. and Levendag, P.C., Ed. (The CyberKnife Society Press, Inc., Sunnyvale, California).

ROGERS, D.W. (1979). "Why not to trust a neutron remmeter," Health Phys. **37**, 735–742.

ROGERS, V.C., NIELSON, K.K. and HOLT, R.B. (1995). "Radon diffusion coefficients for aged residential concretes," Health Phys. **68**(6), 832–834.

SHULTIS, J.K. and FAW, R.E. (1996). *Radiation Shielding* (Prentice Hall PTR, Upper Saddle River, New Jersey).

TAYLOR, P.L., RODGERS, J.E. and SHOBE, J. (1999). "Scatter fractions from linear accelerators with x-ray energies from 6 to 24 MV," Med. Phys. **26**(8), 1442–1446.

THOMAS, D.J, BARDELL A.G. and MACAULAY, E.M. (2002). "Characterisation of a gold foil based Bonner sphere set and measurements of neutron spectra at a medical accelerator," Nucl. Instrum. Methods Phys. Res. A **476**, 31–35.

TOCHILIN, E. and LARIVIERE, P.D. (1979). "Neutron leakage characteristics related to room shielding," in *Proceedings of a Conference on Neutrons from Electron Medical Accelerators*, NBS Special Publication 554, Heaton, H.T. and Jacobs, R., Eds. (U.S. Government Printing Office, Washington).

UWAMINO, Y., NAKAMURA, T., OHKUBO, T. and HARA, A. (1986). "Measurement and calculation of neutron leakage from a medical electron accelerator," Med. Phys. **13**(3), 374–384.

WALKER, R.L. and GROTENHUIS, M. (1961). *A Summary of Shielding Constants for Concrete*, ANL-6443 (Argonne National Laboratory, Argonne, Illinois).

WACHSMANN, F. and DREXLER, G. (1975). *Graphs and Tables for Use in Radiology* (Springer-Verlag, New York).

WU, R.K. and MCGINLEY, P.H. (2003). "Neutron and capture gamma along the mazes of linear accelerator vaults," J. Appl. Clin. Med. Phys. **4**(2), 162–171.

ZAVGORODNI, S.F. (2001). "A method for calculating the dose to a multi-story building due to radiation scattered from the roof of an adjacent radiotherapy facility," Med. Phys. **28**(9), 1926–1930.

The NCRP

The National Council on Radiation Protection and Measurements is a nonprofit corporation chartered by Congress in 1964 to:

1. Collect, analyze, develop and disseminate in the public interest information and recommendations about (a) protection against radiation and (b) radiation measurements, quantities and units, particularly those concerned with radiation protection.
2. Provide a means by which organizations concerned with the scientific and related aspects of radiation protection and of radiation quantities, units and measurements may cooperate for effective utilization of their combined resources, and to stimulate the work of such organizations.
3. Develop basic concepts about radiation quantities, units and measurements, about the application of these concepts, and about radiation protection.
4. Cooperate with the International Commission on Radiological Protection, the International Commission on Radiation Units and Mea- surements, and other national and international organizations, governmental and private, concerned with radiation quantities, units and measurements and with radiation protection.

The Council is the successor to the unincorporated association of scientists known as the National Committee on Radiation Protection and Measurements and was formed to carry on the work begun by the Committee in 1929.

The participants in the Council's work are the Council members and members of scientific and administrative committees. Council members are selected solely on the basis of their scientific expertise and serve as individuals, not as representatives of any particular organization. The scientific committees, composed of experts having detailed knowledge and competence in the particular area of the committee's interest, draft proposed recommendations. These are then submitted to the full membership of the Council for careful review and approval before being published.

The following comprise the current officers and membership of the Council:

Officers

President	Thomas S. Tenforde
Senior Vice President	Kenneth R. Kase
Secretary and Treasurer	David A. Schauer
Assistant Secretary	Michael F. McBride

Members

John F. Ahearne
Sally A. Amundson
Larry E. Anderson
Benjamin R. Archer
Mary M. Austin-Seymour
Steven M. Becker
Joel S. Bedford
Eleanor A. Blakely
William F. Blakely
John D. Boice, Jr.
Wesley E. Bolch
Thomas B. Borak
Andre Bouville
Leslie A. Braby
David J. Brenner
James A. Brink
Antone L. Brooks
Jerrold T. Bushberg
John F. Cardella
Stephanie K. Carlson
Polly Y. Chang
S.Y. Chen
Kelly L. Classic
Mary E. Clark
James E. Cleaver
Michael L. Corradini
J. Donald Cossairt
Allen G. Croff
Francis A. Cucinotta
Paul M. DeLuca
John F. Dicello, Jr.
William P. Dornsife
David A. Eastmond
Stephen A. Feig
Kenneth R. Foster
John R. Frazier
Donald P. Frush
Thomas F. Gesell
Andrew J. Grosovsky
Raymond A. Guilmette
Roger W. Harms
John W. Hirshfeld, Jr.
F. Owen Hoffman
Roger W. Howell
Kenneth R. Kase
Ann R. Kennedy
William E. Kennedy, Jr.
David C. Kocher
Ritsuko Komaki
Amy Kronenberg
Susan M. Langhorst
Howard L. Liber
James C. Lin
Jill A. Lipoti
John B. Little
Paul A. Locke
Jay H. Lubin
C. Douglas Maynard
Claire M. Mays
Cynthia H. McCollough
Barbara J. McNeil
Fred A. Mettler, Jr.
Charles W. Miller
Jack Miller
Kenneth L. Miller
William H. Miller
William F. Morgan
John E. Moulder
David S. Myers
Bruce A. Napier
Carl J. Paperiello
R. Julian Preston
Allan C.B. Richardson
Henry D. Royal
Marvin Rosenstein
Michael T. Ryan
Jonathan M. Samet
Thomas M. Seed
Stephen M. Seltzer
Roy E. Shore
Edward A. Sickles
Steven L. Simon
Paul Slovic
Christopher G. Soares
Daniel J. Strom
Thomas S. Tenforde
Julie E.K. Timins
Lawrence W. Townsend
Lois B. Travis
Robert L. Ullrich
Richard J. Vetter
Daniel E. Wartenberg
Chris G. Whipple
Stuart C. White
J. Frank Wilson
Susan D. Wiltshire
Gayle E. Woloschak
Shiao Y. Woo
Marco A. Zaider
Pasquale D. Zanzonico

Honorary Members

Warren K. Sinclair, *President Emeritus;* Charles B. Meinhold, *President Emeritus*
S. James Adelstein, *Honorary Vice President*
W. Roger Ney, *Executive Director Emeritus*
William M. Beckner, *Executive Director Emeritus*

Seymour Abrahamson
Edward L. Alpen
Lynn R. Anspaugh
John A. Auxier
William J. Bair
Harold L. Beck
Bruce B. Boecker
Victor P. Bond
Robert L. Brent
Reynold F. Brown
Melvin C. Carter
Randall S. Caswell
Frederick P. Cowan
James F. Crow
Gerald D. Dodd
Sarah S. Donaldson
Patricia W. Durbin
Keith F. Eckerman
Thomas S. Ely
Richard F. Foster
R.J. Michael Fry
Ethel S. Gilbert
Joel E. Gray
Robert O. Gorson
Arthur W. Guy
Eric J. Hall
Naomi H. Harley
William R. Hendee
Donald G. Jacobs
Bernd Kahn
Charles E. Land
Roger O. McClellan
Dade W. Moeller
A. Alan Moghissi
Wesley L. Nyborg
John W. Poston, Sr.
Andrew K. Poznanski
Chester R. Richmond
Genevieve S. Roessler
Lawrence N. Rothenberg
Eugene L. Saenger
William J. Schull
John E. Till
Arthur C. Upton
F. Ward Whicker
Marvin C. Ziskin

Lauriston S. Taylor Lecturers

John B. Little (2005) *Nontargeted Effects of Radiation: Implications for Low-Dose Exposures*

Abel J. Gonzalez (2004) *Radiation Protection in the Aftermath of a Terrorist Attack Involving Exposure to Ionizing Radiation*

Charles B. Meinhold (2003) *The Evolution of Radiation Protection: From Erythema to Genetic Risks to Risks of Cancer to ?*

R. Julian Preston (2002) *Developing Mechanistic Data for Incorporation into Cancer Risk Assessment: Old Problems and New Approaches*

Wesley L. Nyborg (2001) *Assuring the Safety of Medical Diagnostic Ultrasound*

S. James Adelstein (2000) *Administered Radioactivity: Unde Venimus Quoque Imus*

Naomi H. Harley (1999) *Back to Background*

Eric J. Hall (1998) *From Chimney Sweeps to Astronauts: Cancer Risks in the Workplace*

William J. Bair (1997) *Radionuclides in the Body: Meeting the Challenge!*

Seymour Abrahamson (1996) *70 Years of Radiation Genetics: Fruit Flies, Mice and Humans*

Albrecht Kellerer (1995) *Certainty and Uncertainty in Radiation Protection*

R.J. Michael Fry (1994) *Mice, Myths and Men*

Warren K. Sinclair (1993) *Science, Radiation Protection and the NCRP*

Edward W. Webster (1992) *Dose and Risk in Diagnostic Radiology: How Big? How Little?*

Victor P. Bond (1991) *When is a Dose Not a Dose?*

J. Newell Stannard (1990) *Radiation Protection and the Internal Emitter Saga*

Arthur C. Upton (1989) *Radiobiology and Radiation Protection: The Past Century and Prospects for the Future*

Bo Lindell (1988) *How Safe is Safe Enough?*

Seymour Jablon (1987) *How to be Quantitative about Radiation Risk Estimates*

Herman P. Schwan (1986) *Biological Effects of Non-ionizing Radiations: Cellular Properties and Interactions*

John H. Harley (1985) *Truth (and Beauty) in Radiation Measurement*

Harald H. Rossi (1984) *Limitation and Assessment in Radiation Protection*

Merril Eisenbud (1983) *The Human Environment—Past, Present and Future*

Eugene L. Saenger (1982) *Ethics, Trade-Offs and Medical Radiation*

James F. Crow (1981) *How Well Can We Assess Genetic Risk? Not Very*

Harold O. Wyckoff (1980) *From "Quantity of Radiation" and "Dose" to "Exposure" and "Absorbed Dose"—An Historical Review*

Hymer L. Friedell (1979) *Radiation Protection—Concepts and Trade Offs*

Sir Edward Pochin (1978) *Why be Quantitative about Radiation Risk Estimates?*

Herbert M. Parker (1977) *The Squares of the Natural Numbers in Radiation Protection*

Currently, the following committees are actively engaged in formulating recommendations:

Program Area Committee 1: Basic Criteria, Epidemiology, Radiobiology, and Risk

SC 1-7 Information Needed to Make Radiation Protection Recommendations for Travel Beyond Low-Earth Orbit

SC 1-8 Risk to Thyroid from Ionizing Radiation

SC 1-13 Impact of Individual Susceptibility and Previous Radiation Exposure on Radiation Risk for Astronauts

SC 1-15 Radiation Safety in NASA Lunar Missions

SC 85 Risk of Lung Cancer from Radon

Program Area Committee 2: Operational Radiation Safety

SC 2-1 Radiation Protection Recommendations for First Responders

SC 46-17 Radiation Protection in Educational Institutions

Program Area Committee 3: Nonionizing Radiation

SC 89-5 Study and Critical Evaluation of Radiofrequency Exposure Guidelines

Program Area Committee 4: Radiation Protection in Medicine

SC 4-1 Management of Persons Contaminated with Radionuclides

SC 91-1 Precautions in the Management of Patients Who Have Received Therapeutic Amounts of Radionuclides

Program Area Committee 5: Environmental Radiation and Radioactive Waste Issues

SC 64-22 Design of Effective Effluent and Environmental Monitoring Programs

SC 64-23 Cesium in the Environment

SC 87-3 Performance Assessment of Near Surface Radioactive Waste Facilities

Program Area Committee 6: Radiation Measurements and Dosimetry

SC 6-1 Uncertainties in the Measurement and Dosimetry of External Radiation Sources

SC 6-2 Radiation Exposure of the U.S. Population

SC 6-3 Uncertainties in Internal Radiation Dosimetry

SC 57-17 Radionuclide Dosimetry Models for Wounds

Advisory Committee 1: Public Policy and Risk Communication

In recognition of its responsibility to facilitate and stimulate cooperation among organizations concerned with the scientific and related aspects of radiation protection and measurement, the Council has created a category of NCRP Collaborating Organizations. Organizations or groups of organizations that are national or international in scope and are concerned with scientific problems involving radiation quantities, units, measurements and effects, or radiation protection may be admitted to collaborating status by the Council. Collaborating Organizations provide a means by which NCRP can gain input into its activities from a wider segment of society. At the same time, the relationships with the Collaborating Organizations facilitate wider dissemination of information about the Council's activities, interests and concerns. Collaborating Organizations have the opportunity to comment on draft reports (at the time that these are submitted to the members of the Council). This is intended to capitalize on the fact that Collaborating Organizations are in an excellent position to both contribute to the identification of what needs to be treated in NCRP reports and to identify problems that might result from proposed recommendations. The present Collaborating Organizations with which NCRP maintains liaison are as follows:

American Academy of Dermatology
American Academy of Environmental Engineers
American Academy of Health Physics
American Association of Physicists in Medicine
American College of Medical Physics
American College of Nuclear Physicians
American College of Occupational and Environmental Medicine
American College of Radiology
American Conference of Governmental Industrial Hygienists
American Dental Association
American Industrial Hygiene Association
American Institute of Ultrasound in Medicine
American Medical Association
American Nuclear Society
American Pharmaceutical Association
American Podiatric Medical Association
American Public Health Association
American Radium Society
American Roentgen Ray Society
American Society for Therapeutic Radiology and Oncology
American Society of Emergency Radiology
American Society of Health-System Pharmacists
American Society of Radiologic Technologists
Association of Educators in Radiological Sciences, Inc.
Association of University Radiologists
Bioelectromagnetics Society
Campus Radiation Safety Officers
College of American Pathologists
Conference of Radiation Control Program Directors, Inc.
Council on Radionuclides and Radiopharmaceuticals
Defense Threat Reduction Agency
Electric Power Research Institute
Federal Communications Commission
Federal Emergency Management Agency
Genetics Society of America
Health Physics Society
Institute of Electrical and Electronics Engineers, Inc.
Institute of Nuclear Power Operations
International Brotherhood of Electrical Workers
National Aeronautics and Space Administration
National Association of Environmental Professionals
National Center for Environmental Health/Agency for Toxic Substances
National Electrical Manufacturers Association
National Institute for Occupational Safety and Health
National Institute of Standards and Technology
Nuclear Energy Institute
Office of Science and Technology Policy
Paper, Allied-Industrial, Chemical and Energy Workers International Union
Product Stewardship Institute
Radiation Research Society
Radiological Society of North America
Society for Risk Analysis
Society of Chairmen of Academic Radiology Departments
Society of Nuclear Medicine
Society of Radiologists in Ultrasound
Society of Skeletal Radiology
U.S. Air Force

U.S. Army
U.S. Coast Guard
U.S. Department of Energy
U.S. Department of Housing and Urban Development
U.S. Department of Labor
U.S. Department of Transportation
U.S. Environmental Protection Agency
U.S. Navy
U.S. Nuclear Regulatory Commission
U.S. Public Health Service
Utility Workers Union of America

NCRP has found its relationships with these organizations to be extremely valuable to continued progress in its program.

Another aspect of the cooperative efforts of NCRP relates to the Special Liaison relationships established with various governmental organizations that have an interest in radiation protection and measurements. This liaison relationship provides: (1) an opportunity for participating organizations to designate an individual to provide liaison between the organization and NCRP; (2) that the individual designated will receive copies of draft NCRP reports (at the time that these are submitted to the members of the Council) with an invitation to comment, but not vote; and (3) that new NCRP efforts might be discussed with liaison individuals as appropriate, so that they might have an opportunity to make suggestions on new studies and related matters. The following organizations participate in the Special Liaison Program:

Australian Radiation Laboratory
Bundesamt fur Strahlenschutz (Germany)
Canadian Nuclear Safety Commission
Central Laboratory for Radiological Protection (Poland)
China Institute for Radiation Protection
Commonwealth Scientific Instrumentation Research Organization (Australia)
European Commission
Health Council of the Netherlands
Institut de Radioprotection et de Surete Nucleaire
International Commission on Non-ionizing Radiation Protection
International Commission on Radiation Units and Measurements
Japan Radiation Council
Korea Institute of Nuclear Safety
National Radiological Protection Board (United Kingdom)
Russian Scientific Commission on Radiation Protection
South African Forum for Radiation Protection
World Association of Nuclear Operations
World Health Organization, Radiation and Environmental Health

NCRP values highly the participation of these organizations in the Special Liaison Program.

The Council also benefits significantly from the relationships established pursuant to the Corporate Sponsor's Program. The program facilitates the interchange of information and ideas and corporate sponsors provide valuable fiscal support for the Council's program. This developing program currently includes the following Corporate Sponsors:

Duke Energy Corporation
GE Healthcare
Global Dosimetry Solutions, Inc.
Landauer, Inc.
Nuclear Energy Institute

The Council's activities have been made possible by the voluntary contribution of time and effort by its members and participants and the generous support of the following organizations:

3M Health Physics Services
Agfa Corporation
Alfred P. Sloan Foundation
Alliance of American Insurers
American Academy of Dermatology
American Academy of Health Physics
American Academy of Oral and Maxillofacial Radiology
American Association of Physicists in Medicine
American Cancer Society
American College of Medical Physics
American College of Nuclear Physicians
American College of Occupational and Environmental Medicine
American College of Radiology
American College of Radiology Foundation
American Dental Association
American Healthcare Radiology Administrators
American Industrial Hygiene Association
American Insurance Services Group
American Medical Association
American Nuclear Society
American Osteopathic College of Radiology
American Podiatric Medical Association
American Public Health Association
American Radium Society
American Roentgen Ray Society
American Society of Radiologic Technologists
American Society for Therapeutic Radiology and Oncology
American Veterinary Medical Association
American Veterinary Radiology Society
Association of Educators in Radiological Sciences, Inc.
Association of University Radiologists
Battelle Memorial Institute
Canberra Industries, Inc.
Chem Nuclear Systems
Center for Devices and Radiological Health
College of American Pathologists
Committee on Interagency Radiation Research and Policy Coordination
Commonwealth Edison
Commonwealth of Pennsylvania
Consolidated Edison
Consumers Power Company
Council on Radionuclides and Radiopharmaceuticals
Defense Nuclear Agency
Defense Threat Reduction Agency
Eastman Kodak Company
Edison Electric Institute
Edward Mallinckrodt, Jr. Foundation
EG&G Idaho, Inc.
Electric Power Research Institute
Electromagnetic Energy Association
Federal Emergency Management Agency
Florida Institute of Phosphate Research
Florida Power Corporation
Fuji Medical Systems, U.S.A., Inc.
Genetics Society of America
Global Dosimetry Solutions
Health Effects Research Foundation (Japan)
Health Physics Society
ICN Biomedicals, Inc.
Institute of Nuclear Power Operations

James Picker Foundation
Martin Marietta Corporation
Motorola Foundation
National Aeronautics and Space Administration
National Association of Photographic Manufacturers
National Cancer Institute
National Electrical Manufacturers Association
National Institute of Standards and Technology
New York Power Authority
Philips Medical Systems
Picker International
Public Service Electric and Gas Company
Radiation Research Society
Radiological Society of North America
Richard Lounsbery Foundation
Sandia National Laboratory
Siemens Medical Systems, Inc.
Society of Nuclear Medicine
Society of Pediatric Radiology
Southern California Edison Company
U.S. Department of Energy
U.S. Department of Labor
U.S. Environmental Protection Agency
U.S. Navy
U.S. Nuclear Regulatory Commission
Victoreen, Inc.
Westinghouse Electric Corporation

Initial funds for publication of NCRP reports were provided by a grant from the James Picker Foundation.

NCRP seeks to promulgate information and recommendations based on leading scientific judgment on matters of radiation protection and measurement and to foster cooperation among organizations concerned with these matters. These efforts are intended to serve the public interest and the Council welcomes comments and suggestions on its reports or activities.

NCRP Publications

NCRP publications can be obtained online in both hard- and soft-copy (downloadable PDF) formats at http://NCRPpublications.org. Professional societies can arrange for discounts for their members by contacting NCRP. Additional information on NCRP publications may be obtained from the NCRP website (http://NCRPonline.org) or by telephone (800-229-2652, ext. 25) and fax (301-907-8768). The mailing address is:

NCRP Publications
7910 Woodmont Avenue
Suite 400
Bethesda, MD 20814-3095

Abstracts of NCRP reports published since 1980, abstracts of all NCRP commentaries, and the text of all NCRP statements are available at the NCRP website. Currently available publications are listed below.

NCRP Reports

No.	Title
8	*Control and Removal of Radioactive Contamination in Laboratories* (1951)
22	*Maximum Permissible Body Burdens and Maximum Permissible Concentrations of Radionuclides in Air and in Water for Occupational Exposure* (1959) [includes Addendum 1 issued in August 1963]
25	*Measurement of Absorbed Dose of Neutrons, and of Mixtures of Neutrons and Gamma Rays* (1961)
27	*Stopping Powers for Use with Cavity Chambers* (1961)
30	*Safe Handling of Radioactive Materials* (1964)
32	*Radiation Protection in Educational Institutions* (1966)
35	*Dental X-Ray Protection* (1970)
36	*Radiation Protection in Veterinary Medicine* (1970)
37	*Precautions in the Management of Patients Who Have Received Therapeutic Amounts of Radionuclides* (1970)
38	*Protection Against Neutron Radiation* (1971)
40	*Protection Against Radiation from Brachytherapy Sources* (1972)
41	*Specification of Gamma-Ray Brachytherapy Sources* (1974)
42	*Radiological Factors Affecting Decision-Making in a Nuclear Attack* (1974)
44	*Krypton-85 in the Atmosphere—Accumulation, Biological Significance, and Control Technology* (1975)
46	*Alpha-Emitting Particles in Lungs* (1975)
47	*Tritium Measurement Techniques* (1976)

49 *Structural Shielding Design and Evaluation for Medical Use of X Rays and Gamma Rays of Energies Up to 10 MeV* (1976)
50 *Environmental Radiation Measurements* (1976)
52 *Cesium-137 from the Environment to Man: Metabolism and Dose* (1977)
54 *Medical Radiation Exposure of Pregnant and Potentially Pregnant Women* (1977)
55 *Protection of the Thyroid Gland in the Event of Releases of Radioiodine* (1977)
57 *Instrumentation and Monitoring Methods for Radiation Protection* (1978)
58 *A Handbook of Radioactivity Measurements Procedures*, 2nd ed. (1985)
60 *Physical, Chemical, and Biological Properties of Radiocerium Relevant to Radiation Protection Guidelines* (1978)
61 *Radiation Safety Training Criteria for Industrial Radiography* (1978)
62 *Tritium in the Environment* (1979)
63 *Tritium and Other Radionuclide Labeled Organic Compounds Incorporated in Genetic Material* (1979)
64 *Influence of Dose and Its Distribution in Time on Dose-Response Relationships for Low-LET Radiations* (1980)
65 *Management of Persons Accidentally Contaminated with Radionuclides* (1980)
67 *Radiofrequency Electromagnetic Fields—Properties, Quantities and Units, Biophysical Interaction, and Measurements* (1981)
68 *Radiation Protection in Pediatric Radiology* (1981)
69 *Dosimetry of X-Ray and Gamma-Ray Beams for Radiation Therapy in the Energy Range 10 keV to 50 MeV* (1981)
70 *Nuclear Medicine—Factors Influencing the Choice and Use of Radionuclides in Diagnosis and Therapy* (1982)
72 *Radiation Protection and Measurement for Low-Voltage Neutron Generators* (1983)
73 *Protection in Nuclear Medicine and Ultrasound Diagnostic Procedures in Children* (1983)
74 *Biological Effects of Ultrasound: Mechanisms and Clinical Implications* (1983)
75 *Iodine-129: Evaluation of Releases from Nuclear Power Generation* (1983)
76 *Radiological Assessment: Predicting the Transport, Bioaccumulation, and Uptake by Man of Radionuclides Released to the Environment* (1984)
77 *Exposures from the Uranium Series with Emphasis on Radon and Its Daughters* (1984)
78 *Evaluation of Occupational and Environmental Exposures to Radon and Radon Daughters in the United States* (1984)
79 *Neutron Contamination from Medical Electron Accelerators* (1984)
80 *Induction of Thyroid Cancer by Ionizing Radiation* (1985)
81 *Carbon-14 in the Environment* (1985)
82 *SI Units in Radiation Protection and Measurements* (1985)
83 *The Experimental Basis for Absorbed-Dose Calculations in Medical Uses of Radionuclides* (1985)
84 *General Concepts for the Dosimetry of Internally Deposited Radionuclides* (1985)

86 *Biological Effects and Exposure Criteria for Radiofrequency Electromagnetic Fields* (1986)
87 *Use of Bioassay Procedures for Assessment of Internal Radionuclide Deposition* (1987)
88 *Radiation Alarms and Access Control Systems* (1986)
89 *Genetic Effects from Internally Deposited Radionuclides* (1987)
90 *Neptunium: Radiation Protection Guidelines* (1988)
92 *Public Radiation Exposure from Nuclear Power Generation in the United States* (1987)
93 *Ionizing Radiation Exposure of the Population of the United States* (1987)
94 *Exposure of the Population in the United States and Canada from Natural Background Radiation* (1987)
95 *Radiation Exposure of the U.S. Population from Consumer Products and Miscellaneous Sources* (1987)
96 *Comparative Carcinogenicity of Ionizing Radiation and Chemicals* (1989)
97 *Measurement of Radon and Radon Daughters in Air* (1988)
99 *Quality Assurance for Diagnostic Imaging* (1988)
100 *Exposure of the U.S. Population from Diagnostic Medical Radiation* (1989)
101 *Exposure of the U.S. Population from Occupational Radiation* (1989)
102 *Medical X-Ray, Electron Beam and Gamma-Ray Protection for Energies Up to 50 MeV (Equipment Design, Performance and Use)* (1989)
103 *Control of Radon in Houses* (1989)
104 *The Relative Biological Effectiveness of Radiations of Different Quality* (1990)
105 *Radiation Protection for Medical and Allied Health Personnel* (1989)
106 *Limit for Exposure to "Hot Particles" on the Skin* (1989)
107 *Implementation of the Principle of As Low As Reasonably Achievable (ALARA) for Medical and Dental Personnel* (1990)
108 *Conceptual Basis for Calculations of Absorbed-Dose Distributions* (1991)
109 *Effects of Ionizing Radiation on Aquatic Organisms* (1991)
110 *Some Aspects of Strontium Radiobiology* (1991)
111 *Developing Radiation Emergency Plans for Academic, Medical or Industrial Facilities* (1991)
112 *Calibration of Survey Instruments Used in Radiation Protection for the Assessment of Ionizing Radiation Fields and Radioactive Surface Contamination* (1991)
113 *Exposure Criteria for Medical Diagnostic Ultrasound: I. Criteria Based on Thermal Mechanisms* (1992)
114 *Maintaining Radiation Protection Records* (1992)
115 *Risk Estimates for Radiation Protection* (1993)
116 *Limitation of Exposure to Ionizing Radiation* (1993)
117 *Research Needs for Radiation Protection* (1993)
118 *Radiation Protection in the Mineral Extraction Industry* (1993)
119 *A Practical Guide to the Determination of Human Exposure to Radiofrequency Fields* (1993)
120 *Dose Control at Nuclear Power Plants* (1994)
121 *Principles and Application of Collective Dose in Radiation Protection* (1995)

122 *Use of Personal Monitors to Estimate Effective Dose Equivalent and Effective Dose to Workers for External Exposure to Low-LET Radiation* (1995)
123 *Screening Models for Releases of Radionuclides to Atmosphere, Surface Water, and Ground* (1996)
124 *Sources and Magnitude of Occupational and Public Exposures from Nuclear Medicine Procedures* (1996)
125 *Deposition, Retention and Dosimetry of Inhaled Radioactive Substances* (1997)
126 *Uncertainties in Fatal Cancer Risk Estimates Used in Radiation Protection* (1997)
127 *Operational Radiation Safety Program* (1998)
128 *Radionuclide Exposure of the Embryo/Fetus* (1998)
129 *Recommended Screening Limits for Contaminated Surface Soil and Review of Factors Relevant to Site-Specific Studies* (1999)
130 *Biological Effects and Exposure Limits for "Hot Particles"* (1999)
131 *Scientific Basis for Evaluating the Risks to Populations from Space Applications of Plutonium* (2001)
132 *Radiation Protection Guidance for Activities in Low-Earth Orbit* (2000)
133 *Radiation Protection for Procedures Performed Outside the Radiology Department* (2000)
134 *Operational Radiation Safety Training* (2000)
135 *Liver Cancer Risk from Internally-Deposited Radionuclides* (2001)
136 *Evaluation of the Linear-Nonthreshold Dose-Response Model for Ionizing Radiation* (2001)
137 *Fluence-Based and Microdosimetric Event-Based Methods for Radiation Protection in Space* (2001)
138 *Management of Terrorist Events Involving Radioactive Material* (2001)
139 *Risk-Based Classification of Radioactive and Hazardous Chemical Wastes* (2002)
140 *Exposure Criteria for Medical Diagnostic Ultrasound: II. Criteria Based on all Known Mechanisms* (2002)
141 *Managing Potentially Radioactive Scrap Metal* (2002)
142 *Operational Radiation Safety Program for Astronauts in Low-Earth Orbit: A Basic Framework* (2002)
143 *Management Techniques for Laboratories and Other Small Institutional Generators to Minimize Off-Site Disposal of Low-Level Radioactive Waste* (2003)
144 *Radiation Protection for Particle Accelerator Facilities* (2003)
145 *Radiation Protection in Dentistry* (2003)
146 *Approaches to Risk Management in Remediation of Radioactively Contaminated Sites* (2004)
147 *Structural Shielding Design for Medical X-Ray Imaging Facilities* (2004)
148 *Radiation Protection in Veterinary Medicine* (2004)
149 *A Guide to Mammography and Other Breast Imaging Procedures* (2004)
150 *Extrapolation of Radiation-Induced Cancer Risks from Nonhuman Experimental Systems to Humans* (2005)
151 *Structural Shielding Design and Evaluation for Megavoltage X- and Gamma-Ray Radiotherapy Facilities* (2005)

Binders for NCRP reports are available. Two sizes make it possible to collect into small binders the "old series" of reports (NCRP Reports Nos. 8–30) and into large binders the more recent publications (NCRP Reports Nos. 32–151). Each binder will accommodate from five to seven reports. The binders carry the identification "NCRP Reports" and come with label holders which permit the user to attach labels showing the reports contained in each binder.

The following bound sets of NCRP reports are also available:

Volume I. NCRP Reports Nos. 8, 22
Volume II. NCRP Reports Nos. 23, 25, 27, 30
Volume III. NCRP Reports Nos. 32, 35, 36, 37
Volume IV. NCRP Reports Nos. 38, 40, 41
Volume V. NCRP Reports Nos. 42, 44, 46
Volume VI. NCRP Reports Nos. 47, 49, 50, 51
Volume VII. NCRP Reports Nos. 52, 53, 54, 55, 57
Volume VIII. NCRP Report No. 58
Volume IX. NCRP Reports Nos. 59, 60, 61, 62, 63
Volume X. NCRP Reports Nos. 64, 65, 66, 67
Volume XI. NCRP Reports Nos. 68, 69, 70, 71, 72
Volume XII. NCRP Reports Nos. 73, 74, 75, 76
Volume XIII. NCRP Reports Nos. 77, 78, 79, 80
Volume XIV. NCRP Reports Nos. 81, 82, 83, 84, 85
Volume XV. NCRP Reports Nos. 86, 87, 88, 89
Volume XVI. NCRP Reports Nos. 90, 91, 92, 93
Volume XVII. NCRP Reports Nos. 94, 95, 96, 97
Volume XVIII. NCRP Reports Nos. 98, 99, 100
Volume XIX. NCRP Reports Nos. 101, 102, 103, 104
Volume XX. NCRP Reports Nos. 105, 106, 107, 108
Volume XXI. NCRP Reports Nos. 109, 110, 111
Volume XXII. NCRP Reports Nos. 112, 113, 114
Volume XXIII. NCRP Reports Nos. 115, 116, 117, 118
Volume XXIV. NCRP Reports Nos. 119, 120, 121, 122
Volume XXV. NCRP Report No. 123I and 123II
Volume XXVI. NCRP Reports Nos. 124, 125, 126, 127
Volume XXVII. NCRP Reports Nos. 128, 129, 130
Volume XXVIII. NCRP Reports Nos. 131, 132, 133
Volume XXIX. NCRP Reports Nos. 134, 135, 136, 137
Volume XXX. NCRP Reports Nos. 138, 139
Volume XXXI. NCRP Report No. 140
Volume XXXII. NCRP Reports Nos. 141, 142, 143
Volume XXXIII. NCRP Report No. 144
Volume XXXIV. NCRP Reports Nos. 145, 146, 147
Volume XXXV. NCRP Reports Nos. 148, 149

(Titles of the individual reports contained in each volume are given previously.)

NCRP Commentaries

No.	Title
1	*Krypton-85 in the Atmosphere—With Specific Reference to the Public Health Significance of the Proposed Controlled Release at Three Mile Island* (1980)

No.	Title
4	*Guidelines for the Release of Waste Water from Nuclear Facilities with Special Reference to the Public Health Significance of the Proposed Release of Treated Waste Waters at Three Mile Island* (1987)
5	*Review of the Publication, Living Without Landfills* (1989)
6	*Radon Exposure of the U.S. Population—Status of the Problem* (1991)
7	*Misadministration of Radioactive Material in Medicine—Scientific Background* (1991)
8	*Uncertainty in NCRP Screening Models Relating to Atmospheric Transport, Deposition and Uptake by Humans* (1993)
9	*Considerations Regarding the Unintended Radiation Exposure of the Embryo, Fetus or Nursing Child* (1994)
10	*Advising the Public about Radiation Emergencies: A Document for Public Comment* (1994)
11	*Dose Limits for Individuals Who Receive Exposure from Radionuclide Therapy Patients* (1995)
12	*Radiation Exposure and High-Altitude Flight* (1995)
13	*An Introduction to Efficacy in Diagnostic Radiology and Nuclear Medicine (Justification of Medical Radiation Exposure)* (1995)
14	*A Guide for Uncertainty Analysis in Dose and Risk Assessments Related to Environmental Contamination* (1996)
15	*Evaluating the Reliability of Biokinetic and Dosimetric Models and Parameters Used to Assess Individual Doses for Risk Assessment Purposes* (1998)
16	*Screening of Humans for Security Purposes Using Ionizing Radiation Scanning Systems* (2003)
17	*Pulsed Fast Neutron Analysis System Used in Security Surveillance* (2003)
18	*Biological Effects of Modulated Radiofrequency Fields* (2003)

Proceedings of the Annual Meeting

No.	Title
1	*Perceptions of Risk*, Proceedings of the Fifteenth Annual Meeting held on March 14-15, 1979 (including Taylor Lecture No. 3) (1980)
3	*Critical Issues in Setting Radiation Dose Limits*, Proceedings of the Seventeenth Annual Meeting held on April 8-9, 1981 (including Taylor Lecture No. 5) (1982)
4	*Radiation Protection and New Medical Diagnostic Approaches,* Proceedings of the Eighteenth Annual Meeting held on April 6-7, 1982 (including Taylor Lecture No. 6) (1983)
5	*Environmental Radioactivity,* Proceedings of the Nineteenth Annual Meeting held on April 6-7, 1983 (including Taylor Lecture No. 7) (1983)
6	*Some Issues Important in Developing Basic Radiation Protection Recommendations*, Proceedings of the Twentieth Annual Meeting held on April 4-5, 1984 (including Taylor Lecture No. 8) (1985)
7	*Radioactive Waste*, Proceedings of the Twenty-first Annual Meeting held on April 3-4, 1985 (including Taylor Lecture No. 9)(1986)
8	*Nonionizing Electromagnetic Radiations and Ultrasound,* Proceedings of the Twenty-second Annual Meeting held on April 2-3, 1986 (including Taylor Lecture No. 10) (1988)

9 *New Dosimetry at Hiroshima and Nagasaki and Its Implications for Risk Estimates*, Proceedings of the Twenty-third Annual Meeting held on April 8-9, 1987 (including Taylor Lecture No. 11) (1988)

10 *Radon*, Proceedings of the Twenty-fourth Annual Meeting held on March 30-31, 1988 (including Taylor Lecture No. 12) (1989)

11 *Radiation Protection Today—The NCRP at Sixty Years*, Proceedings of the Twenty-fifth Annual Meeting held on April 5-6, 1989 (including Taylor Lecture No. 13) (1990)

12 *Health and Ecological Implications of Radioactively Contaminated Environments*, Proceedings of the Twenty-sixth Annual Meeting held on April 4-5, 1990 (including Taylor Lecture No. 14) (1991)

13 *Genes, Cancer and Radiation Protection,* Proceedings of the Twenty-seventh Annual Meeting held on April 3-4, 1991 (including Taylor Lecture No. 15) (1992)

14 *Radiation Protection in Medicine,* Proceedings of the Twenty-eighth Annual Meeting held on April 1-2, 1992 (including Taylor Lecture No. 16) (1993)

15 *Radiation Science and Societal Decision Making,* Proceedings of the Twenty-ninth Annual Meeting held on April 7-8, 1993 (including Taylor Lecture No. 17) (1994)

16 *Extremely-Low-Frequency Electromagnetic Fields: Issues in Biological Effects and Public Health*, Proceedings of the Thirtieth Annual Meeting held on April 6-7, 1994 (not published).

17 *Environmental Dose Reconstruction and Risk Implications,* Proceedings of the Thirty-first Annual Meeting held on April 12-13, 1995 (including Taylor Lecture No. 19) (1996)

18 *Implications of New Data on Radiation Cancer Risk*, Proceedings of the Thirty-second Annual Meeting held on April 3-4, 1996 (including Taylor Lecture No. 20) (1997)

19 *The Effects of Pre- and Postconception Exposure to Radiation*, Proceedings of the Thirty-third Annual Meeting held on April 2-3, 1997, Teratology ***59***, 181–317 (1999)

20 *Cosmic Radiation Exposure of Airline Crews, Passengers and Astronauts*, Proceedings of the Thirty-fourth Annual Meeting held on April 1-2, 1998, Health Phys. ***79***, 466–613 (2000)

21 *Radiation Protection in Medicine: Contemporary Issues*, Proceedings of the Thirty-fifth Annual Meeting held on April 7-8, 1999 (including Taylor Lecture No. 23) (1999)

22 *Ionizing Radiation Science and Protection in the 21st Century*, Proceedings of the Thirty-sixth Annual Meeting held on April 5-6, 2000, Health Phys. ***80***, 317–402 (2001)

23 *Fallout from Atmospheric Nuclear Tests—Impact on Science and Society*, Proceedings of the Thirty-seventh Annual Meeting held on April 4-5, 2001, Health Phys. ***82***, 573–748 (2002)

24 *Where the New Biology Meets Epidemiology: Impact on Radiation Risk Estimates*, Proceedings of the Thirty-eighth Annual Meeting held on April 10-11, 2002, Health Phys. ***85***, 1–108 (2003)

25 *Radiation Protection at the Beginning of the 21st Century–A Look Forward*, Proceedings of the Thirty-ninth Annual Meeting held on April 9–10, 2003, Health Phys. ***87***, 237–319 (2004)

26 *Advances in Consequence Management for Radiological Terrorism Events*, Proceedings of the Fortieth Annual Meeting held on April 14–15, 2004, Health Phys. ***89***, 415–588 (2005)

Lauriston S. Taylor Lectures

No.	Title
1	*The Squares of the Natural Numbers in Radiation Protection* by Herbert M. Parker (1977)
2	*Why be Quantitative about Radiation Risk Estimates?* by Sir Edward Pochin (1978)
3	*Radiation Protection—Concepts and Trade Offs* by Hymer L. Friedell (1979) [available also in *Perceptions of Risk*, see above]
4	*From "Quantity of Radiation" and "Dose" to "Exposure" and "Absorbed Dose"—An Historical Review* by Harold O. Wyckoff (1980)
5	*How Well Can We Assess Genetic Risk? Not Very* by James F. Crow (1981) [available also in *Critical Issues in Setting Radiation Dose Limits*, see above]
6	*Ethics, Trade-offs and Medical Radiation* by Eugene L. Saenger (1982) [available also in *Radiation Protection and New Medical Diagnostic Approaches*, see above]
7	*The Human Environment—Past, Present and Future* by Merril Eisenbud (1983) [available also in *Environmental Radioactivity*, see above]
8	*Limitation and Assessment in Radiation Protection* by Harald H. Rossi (1984) [available also in *Some Issues Important in Developing Basic Radiation Protection Recommendations*, see above]
9	*Truth (and Beauty) in Radiation Measurement* by John H. Harley (1985) [available also in *Radioactive Waste*, see above]
10	*Biological Effects of Non-ionizing Radiations: Cellular Properties and Interactions* by Herman P. Schwan (1987) [available also in *Nonionizing Electromagnetic Radiations and Ultrasound*, see above]
11	*How to be Quantitative about Radiation Risk Estimates* by Seymour Jablon (1988) [available also in *New Dosimetry at Hiroshima and Nagasaki and its Implications for Risk Estimates*, see above]
12	*How Safe is Safe Enough?* by Bo Lindell (1988) [available also in *Radon*, see above]
13	*Radiobiology and Radiation Protection: The Past Century and Prospects for the Future* by Arthur C. Upton (1989) [available also in *Radiation Protection Today*, see above]
14	*Radiation Protection and the Internal Emitter Saga* by J. Newell Stannard (1990) [available also in *Health and Ecological Implications of Radioactively Contaminated Environments*, see above]
15	*When is a Dose Not a Dose?* by Victor P. Bond (1992) [available also in *Genes, Cancer and Radiation Protection*, see above]
16	*Dose and Risk in Diagnostic Radiology: How Big? How Little?* by Edward W. Webster (1992) [available also in *Radiation Protection in Medicine*, see above]
17	*Science, Radiation Protection and the NCRP* by Warren K. Sinclair (1993) [available also in *Radiation Science and Societal Decision Making*, see above]
18	*Mice, Myths and Men* by R.J. Michael Fry (1995)
19	*Certainty and Uncertainty in Radiation Research* by Albrecht M. Kellerer. Health Phys. ***69***, 446–453 (1995)

20 *70 Years of Radiation Genetics: Fruit Flies, Mice and Humans* by Seymour Abrahamson. Health Phys. ***71***, 624–633 (1996)
21 *Radionuclides in the Body: Meeting the Challenge* by William J. Bair. Health Phys. ***73***, 423–432 (1997)
22 *From Chimney Sweeps to Astronauts: Cancer Risks in the Work Place* by Eric J. Hall. Health Phys. ***75***, 357–366 (1998)
23 *Back to Background: Natural Radiation and Radioactivity Exposed* by Naomi H. Harley. Health Phys. ***79***, 121–128 (2000)
24 *Administered Radioactivity: Unde Venimus Quoque Imus* by S. James Adelstein. Health Phys. ***80***, 317–324 (2001)
25 *Assuring the Safety of Medical Diagnostic Ultrasound* by Wesley L. Nyborg. Health Phys. ***82***, 578–587 (2002)
26 *Developing Mechanistic Data for Incorporation into Cancer and Genetic Risk Assessments: Old Problems and New Approaches* by R. Julian Preston. Health Phys. ***85***, 4–12 (2003)
27 *The Evolution of Radiation Protection–From Erythema to Genetic Risks to Risks of Cancer to ?* by Charles B. Meinhold, Health Phys. ***87***, 240–248 (2004)
28 *Radiation Protection in the Aftermath of a Terrorist Attack Involving Exposure to Ionizing Radiation* by Abel J. Gonzalez, Health Phys. ***89***, 418–446 (2005)

Symposium Proceedings

No.	Title
1	*The Control of Exposure of the Public to Ionizing Radiation in the Event of Accident or Attack*, Proceedings of a Symposium held April 27-29, 1981 (1982)
2	*Radioactive and Mixed Waste—Risk as a Basis for Waste Classification,* Proceedings of a Symposium held November 9, 1994 (1995)
3	*Acceptability of Risk from Radiation—Application to Human Space Flight,* Proceedings of a Symposium held May 29, 1996 (1997)
4	*21st Century Biodosimetry: Quantifying the Past and Predicting the Future*, Proceedings of a Symposium held February 22, 2001, Radiat. Prot. Dosim. ***97***(1), (2001)
5	*National Conference on Dose Reduction in CT, with an Emphasis on Pediatric Patients*, Summary of a Symposium held November 6-7, 2002, Am. J. Roentgenol. ***181***(2), 321–339 (2003)

NCRP Statements

No.	Title
1	"Blood Counts, Statement of the National Committee on Radiation Protection," Radiology ***63***, 428 (1954)
2	"Statements on Maximum Permissible Dose from Television Receivers and Maximum Permissible Dose to the Skin of the Whole Body," Am. J. Roentgenol., Radium Ther. and Nucl. Med. ***84***, 152 (1960) and Radiology ***75***, 122 (1960)
3	*X-Ray Protection Standards for Home Television Receivers, Interim Statement of the National Council on Radiation Protection and Measurements* (1968)

4 *Specification of Units of Natural Uranium and Natural Thorium, Statement of the National Council on Radiation Protection and Measurements* (1973)
5 *NCRP Statement on Dose Limit for Neutrons* (1980)
6 *Control of Air Emissions of Radionuclides* (1984)
7 *The Probability That a Particular Malignancy May Have Been Caused by a Specified Irradiation* (1992)
8 *The Application of ALARA for Occupational Exposures* (1999)
9 *Extension of the Skin Dose Limit for Hot Particles to Other External Sources of Skin Irradiation* (2001)
10 *Recent Applications of the NCRP Public Dose Limit Recommendation for Ionizing Radiation* (2004)

Other Documents

The following documents were published outside of the NCRP report, commentary and statement series:

Somatic Radiation Dose for the General Population, Report of the Ad Hoc Committee of the National Council on Radiation Protection and Measurements, 6 May 1959, Science ***131*** (3399), February 19, 482–486 (1960)

Dose Effect Modifying Factors in Radiation Protection, Report of Subcommittee M-4 (Relative Biological Effectiveness) of the National Council on Radiation Protection and Measurements, Report BNL 50073 (T-471) (1967) Brookhaven National Laboratory (National Technical Information Service, Springfield, Virginia)

Residential Radon Exposure and Lung Cancer Risk: Commentary on Cohen's County-Based Study, Health Phys. ***87***(6), 656–658 (2004)

Index